HSP Science

Lab Manual
Grade 4

SCHOOL PUBLISHERS

Visit *The Learning Site!*
www.harcourtschool.com

Printed in the United States of America

ISBN-13: 978-0-15-361003-5
ISBN-10: 0-15-361003-4

2 3 4 5 6 7 8 9 10   073   16 15 14 13 12 11 10 09 08

Contents

Getting Ready for Science

Chapter 1 Classifying Living Things

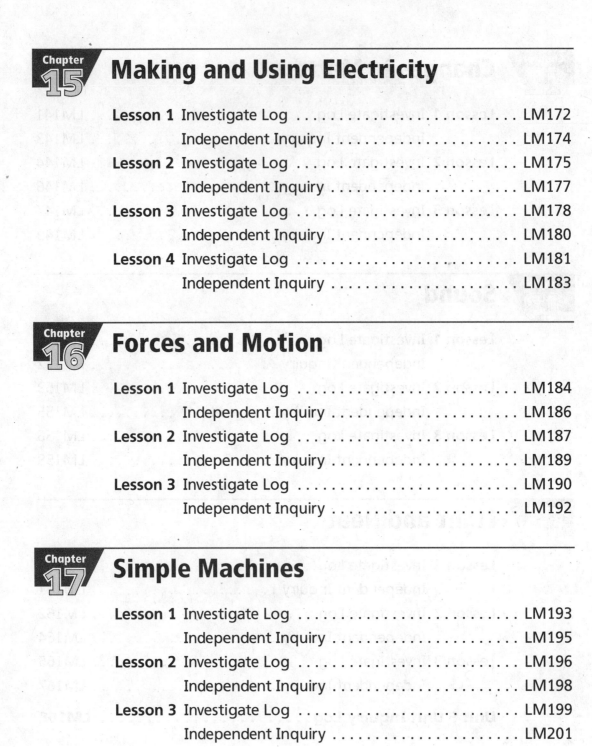

Chapter 15 Making and Using Electricity

Chapter 16 Forces and Motion

Chapter 17 Simple Machines

Safety in Science

Doing investigations in science can be fun, but you need to be sure you do them safely. Here are some rules to follow.

1. **Think ahead.** Study the steps of the investigation so you know what to expect. If you have any questions, ask your teacher. Be sure you understand any caution statements or safety reminders.

2. **Be neat.** Keep your work area clean. If you have long hair, pull it back so it doesn't get in the way. Roll or push up long sleeves to keep them away from your experiment.

3. **Oops!** If you should spill or break something, or get cut, tell your teacher right away.

4. **Watch your eyes.** Wear safety goggles anytime you are directed to do so. If you get anything in your eyes, tell your teacher right away.

5. **Yuck!** Never eat or drink anything during a science activity.

6. **Don't get shocked.** Be especially careful if an electric appliance is used. Be sure that electric cords are in a safe place where you can't trip over them. Don't ever pull a plug out of an outlet by pulling on the cord.

7. **Keep it clean.** Always clean up when you have finished. Put everything away and wipe your work area. Wash your hands.

8. **Play it safe.** Always know where safety equipment, such as fire extinguishers, can be found. Be familiar with how to use the safety equipment around you.

Name _____

Date _____

Science Safety ·

_____ I will study the steps of the investigation before I begin.

_____ I will ask my teacher if I do not understand something.

_____ I will keep my work area clean.

_____ I will pull my hair back and roll up long sleeves before I begin.

_____ I will tell my teacher if I spill or break something or get cut.

_____ I will wear safety goggles when I am told to do so.

_____ I will tell my teacher if I get something in my eye.

_____ I will not eat or drink anything during an investigation unless told to do so by my teacher.

_____ I will be extra careful when using electrical appliances.

_____ I will keep electric cords out of the way and only unplug them by pulling on the protected plug.

_____ I will clean up when I am finished.

_____ I will return unused materials to my teacher.

_____ I will wipe my area and then wash my hands.

Using Science Tools

In almost every science inquiry in this lab manual, you will use tools to observe, measure, and compare the things that you are studying. Each tool is used for a different thing. Part of learning about science is learning how to choose tools to help you answer questions. Here are some tools that you might use and some rules on how to use them.

Using a Hand Lens

Use a hand lens to magnify an object. *Magnify* means "to make something look larger."

1. Hold a hand lens about 12 centimeters (5 inches) from your eye.

2. Bring the object toward you until it comes into focus, which means it is not blurry.

Using a Magnifying Box

Use a magnifying box to magnify something on a flat surface. You can put things inside the box to observe. The magnifying box can be used to observe small living things such as insects. You can observe the insect and then release it.

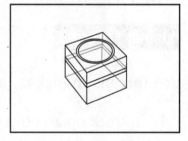

1. Place the magnifying box on top of a flat surface.

2. Place the object you want to observe inside the box. Put on the lid.

3. Look through the lid at the insect or object.

Safety: Do not use a hand lens or magnifying box that is cracked or damaged. If one breaks, do not try to clean up the broken pieces. Call your teacher for help.

Using a Thermometer

Use a thermometer to measure temperature.
Temperature **means "how hot or cold something is."**

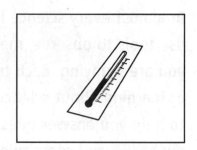

1. Place the thermometer in the liquid. Never stir the liquid with the thermometer. Don't touch the thermometer any more than you need to. When you are measuring the temperature of the air, stand between the thermometer and bright lights, such as sunlight.

2. When you are measuring something that is not being heated or cooled, wait about five minutes for the temperature reading to stop changing. When the thing you are measuring is being heated or cooled, you will not be able to wait. Read the temperature quickly.

3. To read the temperature, move so that your eyes are even with the liquid in the thermometer. Find the scale line that meets the top of the liquid in the thermometer, and read the temperature.

Safety: If a thermometer breaks, call your teacher for help.

Using Forceps

Use forceps to pick up and hold on to objects.

1. To pick up an object, place the tips of the forceps around the object you want to pick up. Press the forceps' handles together with your thumb and first finger to grab the object.

2. Move the object to where you want to place it. Stop pressing with your thumb and first finger. The forceps will let go of the object.

Safety: The tips of forceps can be sharp. Keep forceps away from your face.

Using a Ruler or Meterstick

Use a ruler or a meterstick to measure dimensions of objects and distances. *Dimensions* are the sizes of things, such as *length*, *width*, *height*, and *depth*.

1. Place the zero mark or end of the ruler next to the left end of the object you want to measure.

2. On the other end of the ruler, find the place next to the right end of the object.

3. Look at the scale on the ruler next to the right end of the object. This will show you the length of the object.

4. A meterstick works the same way. You can use it to measure longer objects.

5. You also can use a ruler or meterstick to measure the distance between two places.

Safety: Some rulers and metersticks can have sharp edges. Be careful not to cut yourself. Never point a ruler or meterstick at another person.

Using a Tape Measure

Use a tape measure to measure around curved objects or to measure much longer distances than with a ruler or meterstick.

1. Place the tape measure around the object or along the distance you would like to measure.

2. Read the tape measure the same way you would read a ruler or meterstick.

3. When you are measuring long distances with a tape measure, you will need a lab partner to hold one end while you stretch the tape measure to the other end of the distance.

Using a Balance

Use a balance to measure mass. *Mass* means "how much matter something has."

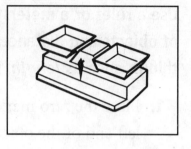

1. Look at the pointer on the base to see that it points at the middle mark of its scale.

2. Place the object that you wish to measure in the pan on the left side of the balance. The pointer should move away from the middle scale mark.

3. Add standard masses to the pan on the right side of the balance. As you add masses, you should see the pointer move back toward the middle scale mark. When the pointer stays at the middle mark, the mass is the same in both pans.

4. Add the numbers on the standard masses that you placed in the pan on the right side of the balance. The total is the mass of the object you measured.

Using a Spring Scale

Use a spring scale to measure an object's weight or a force. *Force* means "how hard you lift or pull on something."

Measuring Weight

1. Hook the spring scale to the object.

2. Lift the scale and object with a smooth motion. Do not jerk the object upward.

3. Wait until the spring stops moving. Then read the force on the scale.

4. You can measure objects that will not fit on the hook. Put the object in a light plastic bag. Hang the bag on the hook.

Using a Measuring Cup, a Beaker, or a Graduate

Use a measuring cup to measure the volume of a liquid or a loose solid such as powder. *Volume* **means "how much space something fills."**

Measuring Liquids

1. Pour the liquid you want to measure into a measuring cup. Put your measuring cup on a flat surface, with the measuring scale facing you.

2. Look at the liquid through the cup. Move so that your eyes are even with the surface of the liquid in the cup. To read the volume of the liquid, find the scale line that is even with the surface of the liquid.

3. When the surface of the liquid is not exactly even with a line, decide which line the liquid is closer to and use that number.

4. Beakers and graduates are used in the same way as measuring cups. They can have different shapes and sizes.

Measuring Solids

- You can measure a solid, such as sand or powder, in the same way that you measure a liquid. You will have to smooth out the top of the solid before you can read the volume.

Safety: Some measuring cups, beakers, and graduates are made of glass. Be careful not to drop one or it could break.

Using a Dropper

Use a dropper to move tiny amounts of liquid.

1. Hold the dropper upright. Put the tip of the dropper in the liquid.

2. Gently squeeze the bulb on the dropper. Stop squeezing to allow the liquid to fill the dropper.

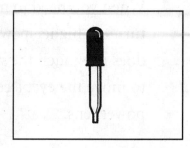

3. Hold the dropper over the place where you want to put the liquid. Gently squeeze the bulb to release one drop at a time.

Safety: Never put a dropper in your mouth. Do not use a dropper that you use in science class for medicine, especially eye drops.

Using a Microscope

Use a microscope to magnify very small objects. A microscope is much more powerful than a simple hand lens.

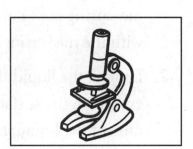

Caring for a Microscope

- Carry a microscope with two hands, use a rolling cart to move it, or ask a teacher for help.

- Never touch any of the lenses of a microscope with your fingers.

Using a Standard Microscope

1. Raise the eyepiece as far as you can by turning the coarse-adjustment knob. Place your slide on the stage.

2. Turn the lowest power lens into place. The lowest power lens is usually the shortest. Slowly use the adjustment knobs to lower the eyepiece and lens to the lowest position it can go without touching the slide.

3. Look through the eyepiece, and raise the eyepiece and lens with the coarse-adjustment knob until the object on the slide is almost in focus. Then use the fine-adjustment knob to focus.

4. When you need to magnify the object on the slide even more, turn the higher-power lens into place. Make sure that the lens does not touch the slide. Use only the fine-adjustment knob to move the eyepiece and lens when looking through a higher-power lens.

Use these pages to plan and conduct a science experiment
to answer a question you may have.

1. Observe and Ask Questions

Make a list of questions you have about a topic. Then
circle a question you want to investigate.

_____

_____

2. Form a Hypothesis

Write a hypothesis. A hypothesis is a scientific explanation that you can test.

_____

3. Plan an Experiment

Identify and Control Variables

To plan your experiment, you must first identify the
important variables. Complete the statements below.

The variable I will change is

_____.

The variables I will observe or measure are

_____.

The variables I will keep the same, or *control*, are

_____.

Develop a Procedure and Gather Materials Write the steps
you will follow to set up an experiment and collect data.

_____

_____

_____

_____

_____

_____

Use extra sheets of blank paper if you need to write down more steps.

Materials List Look carefully at all the steps of your procedure,
and list all the materials you will use. Be sure that your teacher
approves your plan and your materials list before you begin.

_____

_____

_____

_____

_____

_____

4. Conduct the Experiment

Gather and Record Data Follow your plan and collect data. Make a table or chart to record your data. **Observe** carefully. **Record** your observations and be sure to note anything unusual or unexpected. Use the space below and additional paper if necessary.

_____

_____

_____

_____

_____

_____

_____

_____

_____

_____

_____

_____

Name _____

Interpret Data Make a graph of the data you have collected. Plot the data on a sheet of graph paper or use a software program.

5. | **Draw Conclusions and Communicate Results** |

Compare the hypothesis with the data and the graph. Then answer these questions.

1. Given the results of the experiment, do you think the hypothesis is true? Explain.

_____

_____

2. How would you revise the hypothesis? Explain.

_____

_____

3. What else did you **observe** during the experiment?

_____

_____

Prepare a presentation for your classmates to **communicate** what you have learned. Display your data tables and graphs.

Investigate Further

Write another hypothesis that you might investigate.

_____

_____

Name _____

Date _____

Measuring with Straws

Materials

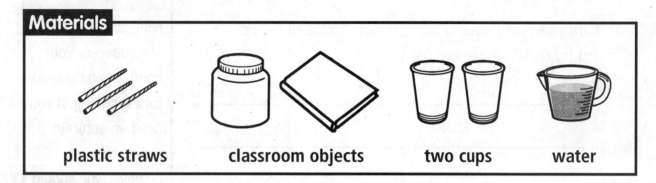

plastic straws   classroom objects   two cups   water

Procedure

1. Use straws to **measure** length and width (distance). For example, you might **measure** this textbook or another flat object. **Record** your measurements.

2. Now use straws to **measure** the distance around a round object (its circumference). Hint: Flatten the straws before you start. **Record** your measurements.

3. Next, think of a way to use straws to **measure** the amount of water in a cup (its volume). Do not share straws. **Record** your measurements.

Object	Measurement

© Harcourt

Draw Conclusions

1. Compare your measurements with those of other students. What can you conclude?

   _____

   _____

   _____

   _____

2. **Inquiry Skill—***Measure* Scientists **measure** carefully so they can record changes accurately. Why do all scientists need to use the same unit of **measurement**?

   _____

   _____

   _____

Inquiry Skill Tip

Nonstandard units, like straws or your hand, can be useful for estimating. If you need an accurate **measurement**, however, you should use standard units such as inches or centimeters on a ruler.

Investigate Self-Assessment	Agree	Not Sure	Disagree
I followed instructions for this investigation.			
I did not share straws with another student.			
I **measured** and recorded the length and width, circumference, or volume of objects.			

© Harcourt

Independent Inquiry

How could you mark a straw to divide it into smaller units?
How would this change the way you collect data? What
might be a reason to do this?

Materials

Here are some materials that you might use.
List additional materials that you need.

- plastic straws   - classroom objects   - two cups   - water

---

1. Explain how you will mark a straw into smaller units.

   _____

   _____

2. Measure the objects from the Investigate again using the
   new units of measurement. Record your observations in the
   table below.

Object	Measurement

3. Explain how the smaller units change the way you collect data.
   What is the purpose of smaller units?

   _____

   _____

Name _____

Date _____

Build a Straw Model

Materials

16 plastic straws   30 paper clips   30 cm masking tape

Procedure

1 You will work with a group to **construct a model** of a building. First, discuss questions such as these: What should our building look like? What are some ways we could use the paper clips and the tape with the straws? How can we keep our building from falling down?

2 Have one group member **record** all your ideas. Be sure to **communicate** well and respect each other's suggestions.

3 **Predict** which techniques will work best, and try them out. **Observe** what works, **draw conclusions**, and **record** them.

4 **Plan** how to construct your building, and then carry out your plan.

Step Number	Ideas and Observations

© Harcourt

Draw Conclusions

1. Why was it important to share ideas before you began construction?

_____

_____

_____

2. **Inquiry Skill**—*Use a Model* Scientists and engineers often **use a model** to better understand how parts work together and to identify problems at the planning stage. What did you learn about constructing a building by making your model?

_____

_____

Investigate Self-Assessment	Agree	Not Sure	Disagree
I discussed with other students ways to build a model.			
I was careful when working with the paper clips.			
I **used a model** to learn about constructing a building.			

Independent Inquiry

Choose one additional material or tool to use in constructing your model. Explain how it will improve your model.

Materials

Here are some materials that you might use.
List additional materials that you need.

- **16 plastic straws**
- **10 paper clips**
- **30 cm masking tape**

1. Identify the additional material or tool you will use. What do you expect it will do?

_____

_____

_____

2. Build the model with the new material or tool. Sketch the model and record your observations.

3. Draw conclusions about how the new material or tool changed or improved the model.

_____

_____

© Harcourt

Testing a Straw Model

Materials

straw models from Lesson 2   paper clips   paper cups   pennies

Procedure

1 Bend paper clips to make a handle for a paper cup, as shown in your textbook.

2 With your group, **predict** how many pennies your straw model can support. Then hang the cup on your model and add one penny at a time. Was your prediction accurate?

3 Now work together to think of ways to strengthen your model. You might also look for other places on your model to hang the cup. **Record** your ideas.

4 **Form a hypothesis** about what will make the model stronger. Then **experiment** to see if the results support each hypothesis.

5 Discuss what made your straw model stronger, and **draw conclusions**.

6 **Communicate** your findings to the class.

Number of Pennies	Result

Draw Conclusions

1. Were you able to increase the strength of your model? How?

_____

_____

_____

_____

2. Inquiry Skill—*Experiment* Scientists **experiment** to test their hypotheses. What did you learn from your experiments in this activity?

_____

_____

> **Inquiry Skill Tip**
>
> Ideas that do not work out are still successes. Scientists and inventors can learn a lot from **experiments** that are "failures." The key is to observe and learn from them.

Investigate Self-Assessment	Agree	Not Sure	Disagree
I hypothesized and counted how many pennies each model could support.			
I was careful when I bent the paper clips.			
I **experimented** with a new model and drew conclusions about what made it stronger.			

© Harcourt

Independent Inquiry

Will your model support more pennies if their weight is spread across the structure? Plan and conduct an experiment to find out.

Materials

Here are some materials that you might use.
List additional materials that you need.

- straw models from Lesson 2
- paper clips
- paper cups
- pennies

1. Describe how you will spread the weight of the pennies across the model.

_____

_____

2. Test your plan and record your observations.

Sketch or Description of Model	Number of Pennies

3. Does your model support more weight if the pennies are spread out? Use the data to justify your answer.

_____

_____

Lung Capacity

1. Observe and Ask Questions

Your lung capacity is how much air your lungs can hold.
Do you have the same lung capacity as other students?
List questions you have about lung capacity. Circle a
question to investigate.

_____

2. Form a Hypothesis

Write a hypothesis. A hypothesis is a scientific explanation
that you can test.

_____

3. Plan an Experiment

Identify and Control Variables To plan your experiment,
you must first identify the important variables. Complete the
statements below.

The variable I will change is _____

The variables I will observe or measure are

_____

The variables I will keep the same, or *control*, are

_____

_____

© Harcourt

Name _____

Develop a Procedure and Gather Materials Write the steps
you will follow to set up an experiment and collect data.

_____

_____

_____

_____

_____

_____

_____

_____

_____

_____

Use extra sheets of blank paper if you need to write down more steps.

Materials List Look carefully at all the steps of your procedure,
and list all the materials you will use. Be sure that your teacher
approves your plan and your materials list before you begin.

_____

_____

_____

_____

Inquiry Log

4. **Conduct the Experiment**

Gather and Record Data Follow your plan and collect data.
Use the chart below or a chart you design to record your
data. **Observe** carefully. **Record** your observations and be
sure to note anything unusual or unexpected.

Name of person	Height of person (cm)	Length of time the pinwheel spun (sec)

© Harcourt

Interpret Data Make graphs of the data you have collected. Plot the data on a sheet of graph paper or use a software program.

5. **Draw Conclusions and Communicate Results**

    Compare the **hypothesis** with the data and the graph. Then answer these questions.

    1. Given the results of the experiment, do you think the hypothesis

       is true? Explain. _____

       _____

    2. How would you revise the hypothesis? Explain.

       _____

       _____

    3. What else did you **observe** during the experiment?

       _____

       _____

Prepare a presentation for your classmates to **communicate** what you have learned. Display your data tables and graphs.

Investigate Further Write another hypothesis that you might investigate.

_____

_____

Name _____

Date _____

Make a Model Cell

Materials

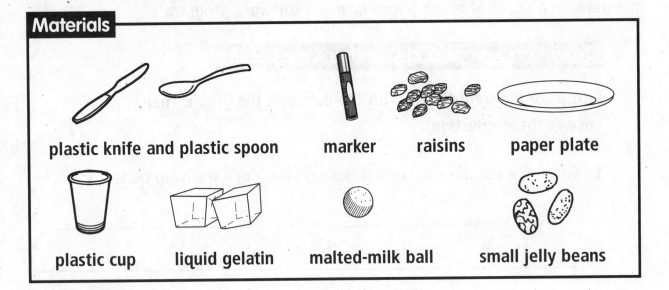

plastic knife and plastic spoon   marker   raisins   paper plate

plastic cup   liquid gelatin   malted-milk ball   small jelly beans

Procedure

1 Write your name on a plastic cup. Then pour liquid gelatin into the plastic cup until it is two-thirds full. Allow the gelatin to set.

2 Carefully remove the gelatin. Use the plastic knife to slice the set gelatin in half. Place the halves on a paper plate.

3 Use a spoon to make a small hole in the center of one of the gelatin halves. Place the malted-milk ball in the hole.

4 Scatter a few raisins and jelly beans on the gelatin that has the malted-milk ball.

5 Place the plain gelatin half on top of the half that contains the cell parts.

6 **Compare** your **model** to the pictures of animal cells in this lesson, which show a cell's parts. List your observations in the chart.

© Harcourt

Cell Model	Cell Pictures

Draw Conclusions

Inquiry Skill Tip

You can **use models** to better understand how things work or to understand the appearance of things. In this investigation, you model an animal cell's appearance. As you observe your model, keep in mind that there are different types of animal cells, and some look very different from the one you have made. Your cell model is useful for helping you understand the basic parts of all animal cells.

1. **Draw conclusions** about what the gelatin, raisins, jellybeans and malted milk ball each represent.

   _____

   _____

2. **Inquiry Skill—***Use Models* Scientists often **use models** to understand complex structures. How does the **model** help you understand some parts of an animal cell?

   _____

   _____

   _____

Investigate Self-Assessment	Agree	Not Sure	Disagree
I followed the directions for making a model cell.			
I obeyed safety rules and didn't put any materials used in the investigation into my mouth.			
I **used the model** to help me understand the parts of a cell.			

Name _____

Date _____

Independent Inquiry

Find out what a bacteria cell looks like. Then make a
model of it. How does this cell compare to the animal cell
you made?

Materials

- plastic knife and plastic spoon
- malted-milk ball
- small jelly beans
- liquid gelatin
- marker
- raisins
- plastic cup
- paper plate

1. Describe the procedure you will follow to make a model of a
   bacteria cell.

   _____

   _____

2. Draw a picture of your bacteria cell model in the space below.

3. How does the bacteria cell compare to the animal cell you made?

   _____

   _____

© Harcourt

Name _____

Date _____

Plant Stems

Materials

blue and red food coloring   water   hand lens   paper towels

carnation with stem   plastic knife   two containers   two clothespins

Procedure

1 Use the plastic knife to trim the end off the carnation stem. Split the stem from the middle to the bottom.

2 Half-fill each container with water. Add blue food coloring to one container, red food coloring to the other container.

3 Place one part of the stem in each container.

4 **Observe** the carnation every 15 minutes. **Record** your observations.

Time (min)	Observations
0	
15	
30	
45	
60	

5 Cut 2 cm off the bottom of each stem part. Use the hand lens to **observe** the cut ends of the stem.

Observations:

_____

© Harcourt

Draw Conclusions

Inquiry Skill Tip

When you **use space relationships** you should also be aware of the sequence of events and how long it takes for things to happen.

1. What do you **observe** about the flower? What do you **observe** about the stem?

   _____

   _____

   _____

2. **Inquiry Skill—*Use Space Relationships*** Scientist can **use space relationships** to better understand what happens in a process. Based on your observations, what can you conclude about the movement of water in stems?

   _____

   _____

   _____

Investigate Self-Assessment	Agree	Not Sure	Disagree
I followed the directions for observing a carnation in colored water.			
I used the hand lens to observe the cut end of the stem.			
I **used space relationships** to sequence the absorption of water.			

© Harcourt

Name _____

Date _____

Independent Inquiry

Use tincture of iodine to test for starch in a carrot.
Where there is starch, the carrot will turn dark. What
can you conclude?

Materials

- tincture of iodine
- plastic knife
- carrot

1. Describe how you will test the carrot.

_____

_____

2. What did you observe when you placed the iodine on the carrot?

_____

_____

_____

3. What can you conclude from your investigation?

_____

_____

_____

© Harcourt

Name _____

Date _____

Backbones

Materials

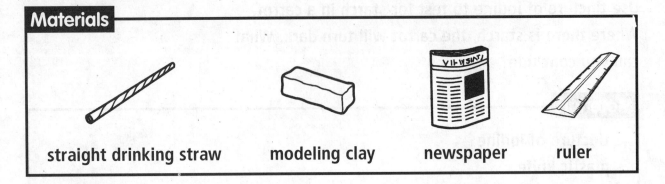

straight drinking straw   modeling clay   newspaper   ruler

Procedure

1 Work in pairs to make **models** to **compare** animals with and without backbones. First, cover your desk with newspaper.

2 Make a base 5 cm in diameter and 3 cm high, using modeling clay.

3 Next, cover a straight drinking straw with modeling clay. The clay should have a diameter of 3 cm to represent the body. The straw represents the backbone.

4 Poke the body into the modeling clay base. You can build the base up around it to support it. The base represents legs. You can add arms and a head, too.

5 Repeat Steps 2–4, but without the straw.

6 **Compare** the way the two models stand. Record the data in the chart.

With Straw	Without Straw

© Harcourt

Draw Conclusions

1. What do you **observe** when you **compare** the two **models**?

_____

_____

_____

2. **Inquiry Skill**—*Plan a Simple Investigation* Scientists **plan simple investigations** to test ideas. What steps would you follow to find out if the thickness of a backbone is important?

_____

_____

Inquiry Skill Tip

When you **plan a simple investigation**, remember to keep track of all of your steps. Others must be able to follow your instructions and duplicate your experiment.

Investigate Self-Assessment	Agree	Not Sure	Disagree
I followed the directions for modeling animals with and without a backbone.			
I used the ruler to measure the diameter and height of the bases.			
I **planned a simple experiment**.			

Name _____

Date _____

Independent Inquiry

Find a garden snail. Observe it. Draw it. Record how it
moves. What can you infer about whether this animal has
a backbone?

Materials

- garden snail
- colored pencils
- paper

1. What will you try to learn in your investigation?

_____

_____

2. Use the table below for your observations.

Observations	Drawing

3. What can you infer about whether or not this animal has a
backbone? Explain.

_____

_____

© Harcourt

Name _____

Date _____

Inherited Characteristics

Materials

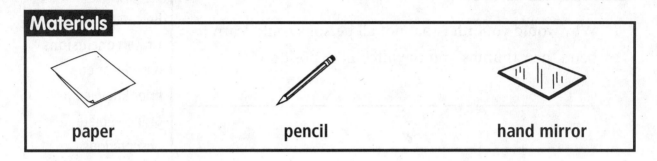

paper   pencil   hand mirror

Procedure

1 Use the chart below for this investigation.

2 Make a fist with your thumb extended, as shown in the picture on page 91 of your textbook. Can you bend your thumb into the hitchhiker's position? **Record** the results.

3 Use the mirror to **observe** your earlobes. Are they attached at the neck, or do they hang free? **Record** your **observations**.

4 Your teacher will ask members of the class to report the results of their **observations**. Tally the results as your classmates report them.

5 Total the number of students who have each trait. Then **calculate** what fraction of the class has each trait.

Trait	Results (Circle one)		Class Totals
Hitchhiker's thumb	Yes	No	
Earlobes	Attached	Free	

© Harcourt

Draw Conclusions

1. Why would you **infer** that not all persons could learn to bend their thumbs into the hitchhiker position?

   _____

   _____

   _____

   _____

2. **Inquiry Skill—*Draw Conclusions*** Based on your **observations** and inferences, what can you **conclude** about hitchhiker thumbs and attached earlobes?

   _____

   _____

Inquiry Skill Tip

Be careful not to **draw conclusions** based just on prior knowledge or incomplete observations. Sometimes prior knowledge may be faulty. If it doesn't agree with your observations, you should always investigate why. You should also make sure your observations are as complete as possible before you **draw conclusions** or you may infer something that is not true.

Investigate Self-Assessment	Agree	Not Sure	Disagree
I followed the directions and tallied the results for this investigation.			
I was careful when handling the hand mirror.			
I **drew conclusions** about hitchhiker thumbs and attached earlobes.			

© Harcourt

Name _____

Date _____

Independent Inquiry

Predict traits your family members will prove to have. Gather data from your family. Use it to complete the chart on this page.

Materials
■ paper   ■ pencil   ■ hand mirror

1. Write your predictions about whether each member of your family will have hitchhiker thumbs and attached earlobes.

   _____

   _____

2. Record data from your observations in the chart below.

Family Member's Name	Hitchhiker Thumbs		Earlobes	
	Yes	No	Attached	Free
	Yes	No	Attached	Free
	Yes	No	Attached	Free
	Yes	No	Attached	Free
	Yes	No	Attached	Free
	Yes	No	Attached	Free

3. What conclusions can you draw from your investigation and data?

   _____

   _____

   _____

© Harcourt

Sprouting Seeds

Materials

radish seeds   water   2 pie pans   2 sponges   cardboard box

Procedure

1 Work in a group. Soak some radish seeds in a cup of water overnight. Write a **hypothesis** about whether seeds need light to germinate.

Hypothesis: _____

2 Place a wet sponge on an aluminum pie pan. Pour about 1 cm of water in the pan.

3 Poke some of the radish seeds into the holes of the sponge.

4 Place the pie pan in a warm, sunny place, and **observe** the plant for the next 3 to 5 days. Be sure the sponge stays moist.

5 Repeat Steps 2 and 3 with the second pie pan. Place the pie pan and sponge on a table and cover with a box so no light can get to it. Be sure the sponge stays wet, but keep it in the dark as much as possible.

6 **Compare** the growth on both sponges and **record** your **observations**.

Day	Plant in Sun	Plant in Dark
1		
2		
3		
4		
5		

Draw Conclusions

1. What do you **conclude** about light and seed germination?

_____

_____

2. Inquiry Skill—*Hypothesize* How did the **experiment** support or reject your **hypothesis**?

_____

_____

_____

_____

Inquiry Skill Tip

Things that you **hypothesize** should always be testable. A hypothesis guides you in the type of experiment you should design. Your hypothesis should mention the variables that you want to investigate.

Investigate Self-Assessment	Agree	Not Sure	Disagree
I followed the directions and recorded observations for this investigation.			
I washed my hands after working with the seeds.			
I **hypothesized** about whether seeds need light to germinate.			

© Harcourt

Name _____

Date _____

Independent Inquiry

Form a hypothesis about whether plants need soil to grow.
For a test, use a radish seed on a moist sponge and a radish
seed in soil in a pot.

Materials

Here are some materials that you might use.
List additional materials that you need.

- radish seeds
- pie pan
- soil
- water
- cardboard box
- sponge
- pot

1. Write your hypothesis about whether plants need soil to grow.

_____

_____

2. Record your observations on the data table below.

Day	Radish Seed Not in Soil	Radish Seed in Soil
1		
2		
3		
4		
5		

3. What conclusion can you draw from your investigation?

_____

_____

© Harcourt

Animal Life Cycles

Materials

| pictures of animal life cycles | paper | pencil | scissors | one paper bag for the whole class |

Procedure

1 Work in a group to understand the life cycle of the animal that is assigned to you.

2 First, study the pictures of the animal's life cycle. Then, draw the stages of the life cycle on a sheet of paper.

3 Cut the paper so that one stage is on each piece. Put the pieces into the bag with your classmates' pieces.

4 Pick a piece from the bag. Then find the students who have the other stages of that animal's life cycle.

5 Put the stages of the cycle in **order**.

6 **Compare** the finished life cycle with the other life cycles. Use the chart to organize your information.

Life Cycle 1	Life Cycle 2

© Harcourt

Draw Conclusions

1. How would you describe animal life cycles in general?

   _____

   _____

   _____

   _____

2. **Inquiry Skill—*Compare*** How did ordering stages in the life cycles and **comparing** them help you describe life cycles?

   _____

   _____

   _____

Inquiry Skill Tip

When you **compare** two different life cycles, start by looking for ways in which they are similar. When you have written down all the ways life cycles are the same, look for ways the two life cycles are different.

Investigate Self-Assessment	Agree	Not Sure	Disagree
I followed the directions and compared cycles for this investigation.			
I was careful when using the scissors.			
I put the stages of the life cycle in **order**.			

© Harcourt

Like a Bird

s

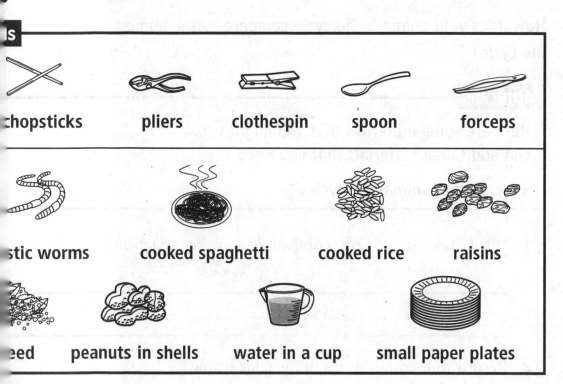

| chopsticks | pliers | clothespin | spoon | forceps |

| stic worms | cooked spaghetti | cooked rice | raisins |

| ed | peanuts in shells | water in a cup | small paper plates |

e

the chart below to record your observations.

"Tools" on one side of your desk, and think of them as
aks. Put each kind of "Food" on a paper plate.

ne type of food in the middle of your desk. Try picking up the
ith each tool (beak), and decide which kind of beak works best.

of the beaks with all of the foods and with the water. Use
rt to **record** your observations and conclusions.

Food	Best Tool (Beak)	Observations

© Harcourt

Independent Inquiry

How does your animal's life cycle compare to the human life cycle?

Materials

Here are some materials that you might use.
List additional materials that you need.

- pictures of animal life cycles

1. What life cycles will you compare in your investigatio[n]

   _____

   _____

2. Record your comparisons in the table below.

Human Life Cycle	My

3. What conclusions can you draw from your comparis[ons]

   _____

   _____

   _____

   _____

Eatin[g

Materi[als

"Tools"	
	2

"Food"

pl[a

birds

Procedu[re

1 Make

2 Put th[e
bird b[

3 Place
food [

4 Test a[
the ch[

LM 44

Draw Conclusions

1. Which kind of beak is best for picking up small seeds? Which kind is best for crushing large seeds?

_____

_____

_____

2. Inquiry Skill—*Draw Conclusions* Scientists experiment and then **draw conclusions** about what they have learned. After experimenting with the tools and foods, what **conclusions** can you draw about why bird beaks are different shapes?

_____

_____

Investigate Self-Assessment	Agree	Not Sure	Disagree
I tested the "bird beaks" according to the directions for this investigation.			
I didn't eat any of the foods that were used in the experiment.			
I **drew a conclusion** about how birds use their beaks to eat.			

Name _____

Date _____

Independent Inquiry

Use a reference book about birds. Match the tools you used with
real bird beaks. Make a hypothesis about how beak shape relates
to food. Then read your book to find out if you are correct.

Materials

Here are some materials that you might use.
List additional materials that you need.

- 2 chopsticks or unsharpened pencils   - clothespin   - spoon
- Reference book about birds   - forceps   - pliers

1. Write your hypothesis about the foods eaten by birds with
different beaks.

_____

_____

2. Use this chart to record your data.

	Hypothesis: How the bird uses its beak	How the bird actually uses its beak
Bird #1		
Bird #2		
Bird #3		
Bird #4		
Bird #5		

3. Did your results match your hypothesis? Explain.

_____

_____

© Harcourt

Train a Fish

Materials

goldfish in a bowl

goldfish food

Procedure

1 Work in a group of three or four to train a goldfish. On the first day, **observe** the behavior of the fish when you hold your hand above the bowl. Then feed the fish by dropping some food into the bowl. **Record** the fish's behavior.

2 The next day, hold your hand a little closer above the bowl. Then drop the pellet. **Record** the fish's behavior.

3 Each day, repeat Step 2, moving your hand nearer the water.

4 Use your **observations** to plan how to continue the **experiment**.

Day	Observations of the Fish's Behavior
1	
2	
3	
4	
5	

Draw Conclusions

1. From your **observations**, what can you **infer** about a fish's ability to learn?

   _____

   _____

   _____

2. **Inquiry Skill—*Observe*** When scientists experiment, they design procedures to gather data. How did **observing** the fish help you plan the experiment?

   _____

   _____

   _____

Inquiry Skill Tip

When you **observed** your fish, were you careful to do so under the same conditions each time? Changing the way in which you observe things can change your observations.

Investigate Self-Assessment	Agree	Not Sure	Disagree
I followed the directions for training the goldfish.			
I did not place my hands in the goldfish bowl, and I washed my hands after the investigation.			
I was able to use observations to plan the **experiment**.			

© Harcourt

Independent Inquiry

**Observe a pet's behavior. List things the pet does that no
one taught it to do. Also list things it may have learned.**

Materials

- pet
- paper
- pencil

1. What will the investigation determine?

_____

_____

2. Classify the behaviors you observe by placing them in the correct
column of the table below.

Behaviors No One Taught the Pet	Behaviors the Pet May Have Learned

3. How were you able to decide if each behavior was taught to the
pet or if it was a behavior that the pet may have learned?

_____

_____

© Harcourt

Name _____

Date _____

Make a Fossil

Materials

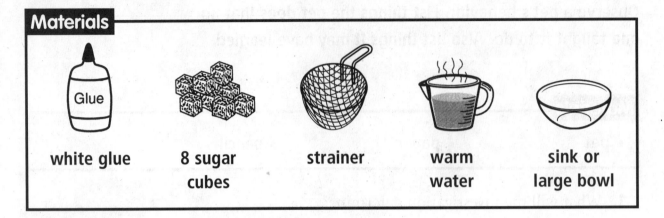

| white glue | 8 sugar cubes | strainer | warm water | sink or large bowl |

Procedure

1 Use the glue to put together 4 sugar cubes, making a 2×2 layer.

2 Glue the other 4 cubes together to make a second layer. Let both layers dry separately for 5 minutes.

3 Spread glue in the shape of a shell on one layer, and place the other layer on top. Let them dry overnight.

4 Put the two-layer structure in the strainer. Hold the strainer over a sink or bowl.

5 Pour warm water over the structure. **Observe** what happens to the sugar and the glue. **Record** your observations.

Observations	
Sugar	**Glue**

Name _____

Draw Conclusions

Inquiry Skill Tip

When you **infer**, think about what you can learn from your observations. After you make an **inference**, think about whether it makes sense. Be sure all of your observations agree with the **inference**.

1. You **made a model** of a fossil. What parts of a plant or an animal did the sugar cubes stand for? What parts did the dried glue stand for?

_____

_____

_____

_____

2. **Inquiry Skill—*Infer*** Scientists **infer**, or explain, based on what they observe. Based on your observations, what can you **infer** about how fossils form?

_____

_____

Investigate Self-Assessment	Agree	Not Sure	Disagree
I followed the directions for making a fossil model.			
I was careful not to place any of the materials used in the investigation into my mouth.			
I **inferred** how fossils form.			

© Harcourt

Name _____

Date _____

Independent Inquiry

Why might fossil skeletons break apart? Hypothesize what will happen if you place your fossil in a bag with some rocks and shake it. Try it. Record your results.

Materials

- **fossil from the Investigate**   ■ **rocks**   ■ **zipper bag**

1. Write a hypothesis for this investigation.

_____

_____

_____

2. Describe your observations.

_____

_____

_____

3. Did the investigation support your hypothesis? Explain.

_____

_____

Sense of Touch

index card   ruler   tape   8 toothpicks

tion: Toothpicks are sharp. Do not play with them. Use only as directed.

edure

ok at the table below. **Predict** which body part has the best
nse of touch. **Record** your prediction.

se a ruler to measure 1 cm along the edge of an index card.
ark it. Tape a toothpick to each of the two marks. The

othpicks should stick about 1 cm past the edge of the card.

peat Step 2 for the other three edges of the index card. Space
e toothpicks 2 cm, 5 cm, and 8 cm apart.

ok away while a partner *lightly* touches each pair of toothpicks
each body part listed. Begin with the 1-cm side, and then use
ch side in turn.

e first time you feel two separate toothpicks, tell your partner.
cord the distance between the toothpicks.

Prediction:		
Distance Apart When Two Toothpicks Are First Felt		
Palm	Lower Arm	Upper Arm
rediction		
ctual		

Pulse Rates

stopwatch, timer, or clock with second hand

Procedure

1. Make a table like the one shown.

2. While you're sitting, find the pulse on your wrist. Count the times
your heart beats in 15 seconds. Multiply the number by 4 to
calculate the number of times your heart beats in a minute while
resting. **Record** the number on your table.

3. Stand up and march in place for 1 minute. As soon as you stop,
find your pulse. Count the times your heart beats in 15 seconds,
and **calculate** the beats per minute. **Record** the number.

4. Rest for a few minutes, and then run in place for 1 minute. As
soon as you stop, find your pulse. Count the times your heart
beats in 15 seconds. Then **calculate** the beats per minute. **Record**
the number.

Activity	Pulse Rate
Sitting	
After marching for 1 minute	
After running for 1 minute	

© Harcourt

© Harcourt

Draw Conclusions

1. Which activity increased your pulse the least? Which
increased it the most?

_____

_____

2. **Inquiry Skills—*Plan an investigation*** Plan an investigation to
find out which activity elevates your heart rate for a longer time.

_____

_____

_____

Investigate Self-Assessment	Agree	Not Sure	Disagree
I measured my pulse according to the directions for this investigation.			
I allowed myself enough space to march and run in place safely.			
I **planned an investigation** about how exercise affects the heart.			

© Harcourt

Independent Inquiry

How do you think exercise affects your breathing rate?
Make a prediction, and design an experiment to find out.

Materials

Here are some materials that you might use.
List additional materials that you need.

- stopwatch, timer, or clock with second hand

1. Describe how you will test the effect of exercise on your
breathing rate. What do you predict your experiment will sho

_____

_____

2. Draw a data table and use it to record your observations.

3. Based on your observations, how do you think your breathin
rate would be affected by even more strenuous exercise? Exp

_____

_____

_____

© Harcourt

Draw Conclusions

1. Which body part felt the two toothpicks the shortest distance apart?

_____

_____

_____

2. **Inquiry Skills—*Infer*** Based on the results of this test, which body part would you **infer** has the best sense of touch? Explain.

_____

_____

_____

> **Inquiry Skill Tip**
>
> When you have carried out an experiment and **inferred** an explanation from your observations, you can test that inference. One experiment leads to another because of inferences.

Investigate Self-Assessment	Agree	Not Sure	Disagree
I tested my sense of touch according to the directions for this investigation.			
I was careful in the way I used the toothpicks.			
I displayed data clearly.			

© Harcourt

Name _____

Date _____

Independent Inquiry

Use your results to predict which will be more sensitive,
your fingertip or the back of your neck. Test your prediction.

Materials

Here are some materials that you might use.
List additional materials that you need.

- index card
- tape
- ruler
- 8 toothpicks

1. Which do you predict is more sensitive, your fingertip or the
   back of your neck?

   _____

   _____

   _____

2. Use this chart to record your data.

	Prediction:	
	Distance Apart When Two Toothpicks Are First Felt	
	fingertip	**back of neck**
Prediction		
Actual		

3. Did your prediction match your results? If one area was more
   sensitive than the other, explain why you think this is so.

   _____

   _____

   _____

Name _____

Date _____

Counting Species

1. Observe and Ask Questions

Every species has a habitat that it prefers to live in. What happens to a plant or animal when humans develop land to build new houses or businesses? Are habitats that have been changed by humans a good place for a lot of plants and animals to live? Make a list of questions you have about species living in habitats that humans have changed. Then circle a question you want to investigate.

_____

2. Form a Hypothesis

Write a hypothesis. A hypothesis is a scientific explanation that you can test.

_____

3. Plan an Experiment

Identify and Control Variables

To plan your experiment, you must first identify the important variables. Complete the statements below.

The variable I will change is

_____

The variables I will observe or measure are

_____

© Harcourt

Name _____

The variables I will keep the same, or *control*, are

_____

_____

Develop a Procedure and Gather Materials Write the steps you
will follow to set up an experiment and collect data.

_____

_____

_____

_____

_____

_____

Use extra sheets of blank paper if you need to write down more steps.

Materials List Look carefully at all the steps of your procedure,
and list all the materials you will use. Be sure that your teacher
approves your plan and your materials list before you begin.

_____

_____

Name ______________________________

Inquiry Log

4. **Conduct the Experiment**

Gather and Record Data Follow your plan and collect data. Use the chart below or a chart you design to record your data. **Observe** carefully. **Record** your observations and be sure to note anything unusual or unexpected.

Habitat Number	Location of Habitat	Description of Habitat	Developed or Natural	Number of Species
1				
2				
3				
4				
5				
6				
7				
8				

Habitat Number	Location of Habitat	Description of Habitat	Developed or Natural	Number of Species
9				
10				
11				
12				
13				
14				
15				
16				

Other Observations

____

____

Interpret Data Make a bar graph of the data you have collected. Plot the data on a sheet of graph paper or use a software program.

5. **Draw Conclusions and Communicate Results**

Compare the hypothesis with the data and the graph. Then answer these questions.

1. Given the results of the experiment, do you think the hypothesis is true? Explain.

_____

_____

2. How would you revise the hypothesis? Explain.

_____

_____

3. What else did you **observe** during the experiment?

_____

Prepare a presentation for your classmates to communicate what you have learned. Display your data tables and graphs.

Investigate Further

Write another hypothesis that you might investigate.

_____

_____

_____

Name _____

Date _____

Modeling an Ecosystem

Materials

gravel

sand

soil

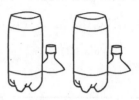
2 empty 2-L soda bottles with the tops cut off

6 small plants

water in a spray bottle

clear plastic wrap

2 rubber bands

Procedure

1 Pour a layer of gravel, then a layer of sand, and then a layer of soil in the bottom of each bottle.

2 Plant three plants in each bottle.

3 Spray the plants and the soil with water. Cover the top of each bottle with plastic wrap. If necessary, hold the wrap in place with a rubber band.

4 Put one of the terrariums you just made in a sunny spot. Put the other one in a dark closet or cabinet.

5 After three days, **observe** each terrarium and **record** what you see.

Plants	Observations
In sun	
In dark	

Name _____

Draw Conclusions

1. What did you **observe** about each of your ecosystems after three days? What part was missing from one ecosystem?

_____

_____

_____

> **Inquiry Skill Tip**
>
> When you **make a model,** avoid including details that make it more complicated to make your observations.

2. Inquiry Skill—*Making a Model* Scientists often learn more about how things affect each other by **making a model.** What did you learn by making a model and observing how its parts interact?

_____

_____

_____

Investigate Self-Assessment	Agree	Not Sure	Disagree
I prepared and observed the terrariums according to the directions for this investigation.			
I washed my hands after handling the gravel, sand, soil, and plants.			
I **made a model** and used it to observe how parts of the model interact.			

© Harcourt

Name _____

Date _____

Pulse Rates

Materials

stopwatch, timer, or clock with second hand

Procedure

1 Make a table like the one shown.

2 While you're sitting, find the pulse on your wrist. Count the times your heart beats in 15 seconds. Multiply the number by 4 to calculate the number of times your heart beats in a minute while resting. **Record** the number on your table.

3 Stand up and march in place for 1 minute. As soon as you stop, find your pulse. Count the times your heart beats in 15 seconds, and **calculate** the beats per minute. **Record** the number.

4 Rest for a few minutes, and then run in place for 1 minute. As soon as you stop, find your pulse. Count the times your heart beats in 15 seconds. Then **calculate** the beats per minute. **Record** the number.

Activity	Pulse Rate
Sitting	
After marching for 1 minute	
After running for 1 minute	

Draw Conclusions

Inquiry Skill Tip

Be sure to follow the steps of the scientific method when you **plan an investigation.**

1. Which activity increased your pulse the least? Which increased it the most?

_____

_____

_____

2. **Inquiry Skills—*Plan an investigation*** Plan an investigation to find out which activity elevates your heart rate for a longer time.

_____

_____

_____

Investigate Self-Assessment	Agree	Not Sure	Disagree
I measured my pulse according to the directions for this investigation.			
I allowed myself enough space to march and run in place safely.			
I **planned an investigation** about how exercise affects the heart.			

© Harcourt

Investigate Log

Independent Inquiry

How do you think exercise affects your breathing rate?
Make a prediction, and design an experiment to find out.

Materials

Here are some materials that you might use.
List additional materials that you need.

- stopwatch, timer, or clock with second hand

---

1. Describe how you will test the effect of exercise on your breathing rate. What do you predict your experiment will show?

   _____

   _____

   _____

   _____

2. Draw a data table and use it to record your observations.

3. Based on your observations, how do you think your breathing rate would be affected by even more strenuous exercise? Explain.

   _____

   _____

   _____

© Harcourt

Name _____

Date _____

The Sense of Touch

Materials

index card

ruler

tape

8 toothpicks

Caution: Toothpicks are sharp. Do not play with them. Use only as directed.

Procedure

1 Look at the table below. **Predict** which body part has the best sense of touch. **Record** your prediction.

2 Use a ruler to measure 1 cm along the edge of an index card. Mark it. Tape a toothpick to each of the two marks. The toothpicks should stick about 1 cm past the edge of the card.

3 Repeat Step 2 for the other three edges of the index card. Space the toothpicks 2 cm, 5 cm, and 8 cm apart.

4 Look away while a partner *lightly* touches each pair of toothpicks to each body part listed. Begin with the 1-cm side, and then use each side in turn.

5 The first time you feel two separate toothpicks, tell your partner. **Record** the distance between the toothpicks.

	Prediction:		
	Distance Apart When Two Toothpicks Are First Felt		
	Palm	Lower Arm	Upper Arm
Prediction			
Actual			

© Harcourt

Name _____

Draw Conclusions

1. Which body part felt the two toothpicks the shortest distance apart?

_____

_____

_____

2. Inquiry Skills—*Infer* Based on the results of this test, which body part would you **infer** has the best sense of touch? Explain.

_____

_____

_____

Inquiry Skill Tip

When you have carried out an experiment and **inferred** an explanation from your observations, you can test that inference. One experiment leads to another because of inferences.

Investigate Self-Assessment	Agree	Not Sure	Disagree
I tested my sense of touch according to the directions for this investigation.			
I was careful in the way I used the toothpicks.			
I displayed data clearly.			

© Harcourt

Name _____

Date _____

Independent Inquiry

Use your results to predict which will be more sensitive,
your fingertip or the back of your neck. Test your prediction.

Materials

Here are some materials that you might use.
List additional materials that you need.

- index card
- tape
- ruler
- 8 toothpicks

---

1. Which do you predict is more sensitive, your fingertip or the
back of your neck?

_____

_____

_____

2. Use this chart to record your data.

	Prediction:	
	Distance Apart When Two Toothpicks Are First Felt	
	fingertip	back of neck
Prediction		
Actual		

3. Did your prediction match your results? If one area was more
sensitive than the other, explain why you think this is so.

_____

_____

_____

Name _____

Date _____

Counting Species

1. Observe and Ask Questions

Every species has a habitat that it prefers to live in. What happens to a plant or animal when humans develop land to build new houses or businesses? Are habitats that have been changed by humans a good place for a lot of plants and animals to live? Make a list of questions you have about species living in habitats that humans have changed. Then circle a question you want to investigate.

_____

2. Form a Hypothesis

Write a hypothesis. A hypothesis is a scientific explanation that you can test.

_____

3. Plan an Experiment

Identify and Control Variables

To plan your experiment, you must first identify the important variables. Complete the statements below.

The variable I will change is

_____

The variables I will observe or measure are

_____

© Harcourt

The variables I will keep the same, or *control*, are

_____

_____

Develop a Procedure and Gather Materials Write the steps you
will follow to set up an experiment and collect data.

_____

_____

_____

_____

_____

_____

Use extra sheets of blank paper if you need to write down more steps.

Materials List Look carefully at all the steps of your procedure,
and list all the materials you will use. Be sure that your teacher
approves your plan and your materials list before you begin.

_____

_____

Name _____

4. **Conduct the Experiment**

Gather and Record Data Follow your plan and collect data. Use the chart below or a chart you design to record your data. **Observe** carefully. **Record** your observations and be sure to note anything unusual or unexpected.

Habitat Number	Location of Habitat	Description of Habitat	Developed or Natural	Number of Species
1				
2				
3				
4				
5				
6				
7				
8				

Habitat Number	Location of Habitat	Description of Habitat	Developed or Natural	Number of Species
9				
10				
11				
12				
13				
14				
15				
16				

Other Observations

_____

_____

Interpret Data Make a bar graph of the data you have collected.
Plot the data on a sheet of graph paper or use a software program.

5. **Draw Conclusions and Communicate Results**

Compare the hypothesis with the data and the graph. Then
answer these questions.

1. Given the results of the experiment, do you think the hypothesis
   is true? Explain.

   _____

   _____

2. How would you revise the hypothesis? Explain.

   _____

   _____

3. What else did you **observe** during the experiment?

   _____

Prepare a presentation for your classmates to communicate what
you have learned. Display your data tables and graphs.

Investigate Further

Write another hypothesis that you might investigate.

_____

_____

_____

Name _____

Date _____

Modeling an Ecosystem

Materials

gravel

sand

soil

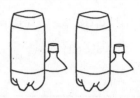

2 empty 2-L soda bottles
with the tops cut off

6 small plants

water in a spray bottle

clear plastic wrap

2 rubber bands

Procedure

1 Pour a layer of gravel, then a layer of sand, and then a layer of soil
in the bottom of each bottle.

2 Plant three plants in each bottle.

3 Spray the plants and the soil with water. Cover the top of each
bottle with plastic wrap. If necessary, hold the wrap in place with
a rubber band.

4 Put one of the terrariums you just made in a sunny spot. Put the
other one in a dark closet or cabinet.

5 After three days, **observe** each terrarium and **record** what you see.

Plants	Observations
In sun	
In dark	

© Harcourt

Use with pages 200–201.   (page 1 of 3)   **Lab Manual**

Name _____

Draw Conclusions

Inquiry Skill Tip

When you **make a model,** avoid including details that make it more complicated to make your observations.

1. What did you **observe** about each of your ecosystems after three days? What part was missing from one ecosystem?

_____

_____

_____

2. **Inquiry Skill—*Making a Model*** Scientists often learn more about how things affect each other by **making a model.** What did you learn by making a model and observing how its parts interact?

_____

_____

_____

Investigate Self-Assessment	Agree	Not Sure	Disagree
I prepared and observed the terrariums according to the directions for this investigation.			
I washed my hands after handling the gravel, sand, soil, and plants.			
I **made a model** and used it to observe how parts of the model interact.			

© Harcourt

Name _____

Date _____

Independent Inquiry

What effect does sunlight have on seeds that have just been planted? First, write your hypothesis. Then plan an experiment to see if your hypothesis is supported.

Materials

Here are some materials that you might use.
List additional materials that you need.

- gravel
- sand
- soil
- water in a spray bottle
- clear plastic wrap
- 2 empty 2-L soda bottles with the tops cut off
- 2 rubber bands
- seeds

1. What do you hypothesize will happen to the seeds?

_____

_____

2. Use this chart to record your data.

Day	Seed Observations
1	
2	
3	
4	
5	
6	
7	

3. Did your observations support your hypothesis? Explain.

_____

_____

Name _____

Date _____

Observing the Effects of Water

Materials

4 small identical plants in clay pots

large labels

water

Procedure

1 Use the labels to number the pots 1, 2, 3, and 4. Label pots 1 and 2 *watered*. Label pots 3 and 4 *not watered*.

2 Make a table like the one shown here. Draw a picture of each plant under Day 1.

3 Place all four pots in a sunny window. Water all four pots until the soil is a little moist. Keep the soil of pots 1 and 2 moist during the whole experiment. Do not water pots 3 and 4 again.

4 Wait three days. Then **observe** and **record** how each plant looks. Draw a picture of each one under Day 4.

5 Repeat Step 4 twice. Draw pictures of the plants on Days 7 and 10.

	Day 1	Day 4	Day 7	Day 10
Plant 1 (watered)				
Plant 2 (watered)				
Plant 3 (not watered)				
Plant 4 (not watered)				

© Harcourt

Draw Conclusions

1. What changes did you **observe** during this Investigate? What do they tell you?

   _____

   _____

   _____

2. **Inquiry Skill—*Compare*** Scientists **compare** changes to determine how one thing affects another. How could you **compare** how fast the soil dries out in a clay pot and a plastic pot?

   _____

   _____

   _____

Investigate Self-Assessment	Agree	Not Sure	Disagree
I prepared and observed the pots according to the directions for this investigation.			
I washed my hands after handling the plants and the soil.			
I **compared** how the plants responded to different conditions.			

Name _____

Date _____

Independent Inquiry

How does covering a plant with plastic wrap affect the
plant's need for water? Write your hypothesis. Then design
and carry out an experiment to check your hypothesis.

Materials

Here are some materials that you might use.
List additional materials that you need.

- 4 small identical plants in clay pots
- large labels

- water
- plastic wrap

1. State your hypothesis for this investigation and your plan for an
   experiment to test the hypothesis.

_____

2. Use this chart to record your observations.

	Day 1	Day 2	Day 3	Day 4
Plant 1 (covered)				
Plant 2 (covered)				
Plant 3 (uncovered)				
Plant 4 (uncovered)				

3. Did your experiment confirm your hypothesis? Explain your
   results.

_____

Name _____

Date _____

Losing It: Observing Erosion

Materials

marker   2 paper cups   measuring cup   water

2 clean plastic foam trays   soil and small rocks   sharpened pencil

Procedure

1 Use the marker to write *A* on one tray and on one paper cup. Write *B* on the second tray and on the second paper cup.

2 In each tray, make an identical slope out of soil and rocks.

3 Carefully use the pencil to make three small holes in the bottom of cup A. Make six larger holes in the bottom of cup B.

4 **Record** how the two slopes look now. Label your drawings *A* and *B*.

5 Hold cup A over the slope in tray A. Slowly pour 1 cup of water into the paper cup, and let it run down the slope. **Record** how the slope looks now.

6 Repeat Step 5 with cup B and tray B. Then **record** how the slope looks.

	Before being watered	After being watered
Tray A Cup A		
Tray B Cup B		

© Harcourt

Draw Conclusions

1. Compare both slopes at the end of the activity. What did each cup represent?

   _____

   _____

   _____

Inquiry Skill Tip

Think about your **experiment**. Pay attention to what you control and what you change. Change only one thing at a time.

2. **Inquiry Skill—***Experiment* An **experiment** is a careful, controlled test. What were you testing and what did you control in the activity?

   _____

   _____

Investigate Self-Assessment	Agree	Not Sure	Disagree
I prepared and tested the slopes according to the directions for this investigation.			
I used the sharpened pencil in a careful way.			
I made an **inference** about the difference between the two slopes.			

© Harcourt

Name _____

Date _____

Independent Inquiry

Try the same activity, using only rocks, using level soil, or using plants growing in the soil. Make a prediction about what will happen, and then do the activity to see if your prediction was accurate.

Materials

Here are some materials that you might use.
List additional materials that you need.

- 2 paper cups
- sharpened pencil
- 2 clean plastic foam trays
- measuring cup
- small rocks
- plants
- water
- marker
- soil

1. Describe your experimental set-up and what you predict will happen under the different conditions.

_____

_____

2. Use this space to record your observations.

	Before Being Watered	After Being Watered
Rocks		
Level soil		
Soil with plants		

3. How did your results compare with your predictions?

_____

_____

© Harcourt

Name _____

Date _____

Decomposing Bananas

Materials

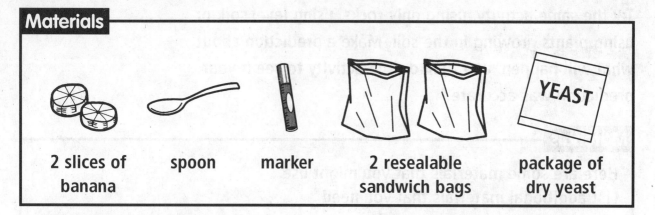

| 2 slices of banana | spoon | marker | 2 resealable sandwich bags | package of dry yeast |

Procedure

1 Put a banana slice in each bag.

2 Sprinkle $\frac{2}{3}$ spoonful of dry yeast on one banana slice. Yeast is a decomposer, so use the marker to label this bag *D*.

3 Close both bags. Put the bags in the same place.

4 Check both bags every day for a week. **Observe** and **record** the changes you see in each bag.

Day	Banana Without Yeast	Banana With Yeast
1		
2		
3		
4		
5		

Draw Conclusions

1. Which banana slice shows more changes? What is the cause of these changes?

   _____

   _____

   _____

2. **Inquiry Skill—*Use Time Relationships*** Scientists **use time relationships** to measure progress. How long did it take for your banana slices to begin showing signs of decomposition? How long do you think it would take for your banana slices to completely decompose?

   _____

   _____

   _____

Inquiry Skill Tip

By thinking about what causes the changes, you can infer other properties of the things. You **use time relationships** to understand processes. You need to watch carefully for small changes. You also need to keep careful track of time. It's best to observe after equal amounts of time, for example, every day or every five minutes. Then by thinking about changes and time, you can infer an explanation of a process.

Investigate Self-Assessment	Agree	Not Sure	Disagree
I followed the directions for observing changes in the decomposing bananas.			
I obeyed safety rules by not placing any of the materials used in the investigation in my mouth.			
Based on my comparisons of the samples, I was able to **use time relationships** to infer something about the yeast.			

© Harcour

Name _____

Date _____

Independent Inquiry

What would happen if you put flour on one banana slice
instead of yeast? Write down your prediction and then try it.

Materials

- **2 slices of banana**
- **2 resealable sandwich bags**
- **marker**
- **spoon**
- **flour**

1. Write your prediction.

_____

_____

2. Record your observations on the table below.

Day	Banana Without Flour	Banana With Flour
1		
2		
3		
4		
5		

3. Explain how the results of this investigation were the same or
different from your prediction.

_____

_____

_____

Name _____

Date _____

Make a Food Chain

Materials

8–10 blank index cards

reference books about animals

colored pencils or markers

Procedure

1 Choose a place where animals live. Some examples are a pine forest, rain forest, desert, wetlands, and the ocean.

2 On an index card, draw an animal that lives in the place you have chosen. Draw more animals, one kind per card. Include large animals and small animals. Draw producers on some of the cards. Look up information about plants and animals if you need help.

3 Put your cards in **order** by what eats what. You might have more than one set of cards. If one of your animals does not fit anywhere, trade cards with someone. You can also draw an animal to link two of your cards. For example, you could draw a rabbit to link a grass card and a hawk card. Draw your finished food chain below.

Your Food Chain	Other Food Chain

© Harcourt

Draw Conclusions

1. Could the same animal fit into more than one set of cards? Explain your answer.

   _____

   _____

   _____

2. **Inquiry Skill—*Communicate*** Scientists **communicate** their ideas in many ways. What do your cards **communicate** about the relationship of these living things to one another?

   _____

   _____

Inquiry Skill Tip

Pictures are a good way to **communicate** information. They help you see much more than can be communicated by words. The cards you use in this investigation help you organize the animals by communicating many of their characteristics.

Investigate Self-Assessment	Agree	Not Sure	Disagree
I followed the directions for making a food chain.			
I used colored pencils or markers to draw animals on the cards.			
I was able to **communicate** about the relationship of the animals to one another.			

© Harcourt

Independent Inquiry

Draw a series of cards in order, with yourself as the last consumer. Compare your role of consumer with the roles of other consumers.

Materials

- 8–10 blank index cards
- colored pencils or markers

1. What is the purpose of this investigation?

_____

_____

2. Write an ordered list of the names of the producers and consumers that you drew.

_____

_____

3. How does your role of consumer compare with the roles of other consumers?

_____

_____

_____

_____

Earthquake-Resistant Structures

1. **Observe and Ask Questions**

How can you minimize earthquake damage in your home? List questions you have about earthquake damage to buildings. Circle a question you want to investigate.

_____

_____

2. **Form a Hypothesis**

Write a hypothesis. A hypothesis is a scientific explanation that can be tested.

_____

3. **Plan an Experiment**

Identify and Control Variables To plan your experiment, you must first identify the important variables. Complete the statements below.

The variable I will change is

_____

The variables I will observe or measure are

_____

The variables I will keep the same, or *control*, are

_____

_____

© Harcourt

Develop a Procedure and Gather Materials Write the steps you
will follow to set up an experiment and collect data.

_____

_____

_____

_____

_____

_____

_____

_____

_____

Use extra sheets of blank paper if you need to write down more steps.

Materials List Look carefully at all the steps of your procedure,
and list all the materials you will use. Be sure that your teacher
approves your plan and your materials list before you begin.

_____

_____

_____

© Harcourt

4. Conduct the Experiment

Gather and Record Data Follow your plan and collect data.
Use the chart below or a chart you design to record your
data. **Observe** carefully. **Record** your observations and be
sure to note anything unusual or unexpected. Draw the inside
of your model house before and after each earthquake.

House #1

Before the Earthquake   After the Earthquake

House #2

Before the Earthquake   After the Earthquake

Observations

_____

_____

5. Draw Conclusions and Communicate Results

Compare the hypothesis with the data. Then answer these questions.

1. Given the results of the experiment, do you think the hypothesis is true? Explain.

_____

2. How would you revise the hypothesis? Explain.

_____

3. What else did you **observe** during the experiment?

_____

_____

_____

Prepare a presentation for your classmates to communicate what you have learned. Display your data and pictures.

Investigate Further

Write another hypothesis that you might investigate.

_____

_____

Name _____

Date _____

Making Sedimentary Rock

Materials

2 small plastic cups   sand   measuring cup   white glue   water

plastic stirrer   scissors   pushpin   1 large plastic cup   hand lens

Procedure

1 Use the pushpin to make a small hole in the bottom of one cup. The hole should be big enough to let water out but not sand.

2 Place 60 mL water and 60 mL white glue in the other cup. Mix and set aside.

3 Fill the first cup with sand.

4 Using a ring stand, suspend the cup holding sand over the pan.

5 Pour the glue mixture into the sand. Let the liquid drain into the pan. Let the glue dry. This could take two or three days.

6 When the liquid stops draining, remove the cup with the sand in it. Cut away the plastic cup with the scissors. Record your observations about your results in the chart.

Step	Observations

© Harcourt

Draw Conclusions

1. **Observe** and describe the structure that has formed in the cup. Use the hand lens.

   _____

   _____

   _____

2. **Inquiry Skill—*Compare*** Compare the way you made your rock with the way you think an actual rock would form. Check your answer when you finish the lesson.

   _____

   _____

   _____

   _____

Investigate Self-Assessment	Agree	Not Sure	Disagree
I followed the directions for making a model of sedimentary rock.			
I was careful when using the scissors.			
I **used the model** of sedimentary rock to learn how an actual rock would form.			

© Harcourt

Name _____

Date _____

Independent Inquiry

Make a sandstone rock with several layers. Plan the investigation so you can easily observe each layer in the rock.

Materials

Here are some materials that you might use.
List additional materials that you need.

- plastic stirrer for mixing
- small plastic cup
- aluminum pan
- measuring cup

- ring stand
- hand lens
- white glue

- pushpin
- sand
- water

1. Describe how you will make a sandstone rock with layers.

_____

2. Observe the finished rock model. Draw and label a picture to show it.

3. Explain how the model is like a real piece of sandstone. Explain how it is different.

_____

_____

_____

© Harcourt

Model a Rock Cycle

Materials

small plastic pencil sharpener

crayons of three colors

metal cookie sheet

waxed paper

iron

aluminum pie pan

toaster oven

Procedure

1 Use the sharpener to make three piles of crayon shavings, each a different color.

2 Place the crayon shavings in three layers, on the cookie sheet. Press down the layers with your hand. Draw what you **observe**.

3 Place the waxed paper over the shavings. Your teacher will press down on the shavings lightly with a warm iron. The teacher will leave the iron for a few seconds, until the shavings start to melt. They should not melt completely. Let the shavings cool for a couple of minutes. Draw what you **observe**.

4 Place the block of shavings into the pie pan. **CAUTION:** Your teacher will put the pan in the toaster oven. Let the shavings melt. Your teacher will remove the shavings and let them cool. Record the layers of your model on the chart.

Material	Result
Step 2	
Step 3	
Step 4	

© Harcourt

Draw Conclusions

1. What type of rock does Step 2 represent? What type of rock does Step 3 represent? How about Step 4?

   _____

   _____

2. Inquiry Skill—*Plan and Conduct a Simple Investigation* How would you **plan and conduct a simple investigation** that uses the "rock" from Step 4 to **model** how sedimentary rock forms.

   _____

   _____

   _____

> **Inquiry Skill Tip**
>
> A **simple investigation** should be short and quick. Its goal is to observe a response or to show how a thing works. To plan an investigation, write down what you will do. Then list the materials you will need.

Investigate Self-Assessment	Agree	Not Sure	Disagree
I followed the directions to model different stages of the rock cycle.			
I obeyed all safety rules when working with hot appliances.			
I **used models** to represent different types of rocks.			

© Harcourt

Independent Inquiry

Design a similar investigation that models the same cycle
but with the events in a different sequence.

Materials
■ small plastic pencil sharpener   ■ waxed paper   ■ iron
■ crayons of three colors   ■ aluminum pie pan   ■ toaster oven
■ metal cookie sheet

1. List the steps you will follow in your investigation to model the
   rock cycle in a different sequence of events.

   _____

   _____

2. Make a diagram to show the sequence of events in the rock cycle
   that your model represents.

3. Compare this rock cycle to the rock cycle you made in the
   Investigate. How are they similar and how are they different?

   _____

   _____

Name _____

Date _____

 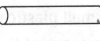

Shake Things Up

Materials

6 medium-size rocks

empty, clear plastic juice container with lid (2 qts/1.89 L)

2 pieces of chalk

Procedure

1 **Make a model** of the way rocks break down in nature. Add two pieces of chalk to the container.

2 Place six rocks in a container.

3 Put the lid on the container.

4 Shake the container so that the rocks and chalk rub against each other. Do this for several minutes. You can take turns with your lab partner.

Observations	
Start	
After 1 minute	
After 2 minutes	
After 3 minutes	
After 4 minutes	

© Harcourt

Draw Conclusions

1. **Compare** the way the rocks and chalk looked at the start and at the end of the investigation.

_____

_____

2. **Inquiry Skill—***Infer* Scientists often **infer** the reasons for an investigation's results. Why did some of the materials in the container break down faster than other materials? How do you think this relates to rocks in nature?

_____

_____

_____

Inquiry Skill Tip

When you use a model to **infer**, be sure that the model represents actual objects and events in nature. *Ask:* Do these materials behave the same way as the real-life objects? Did my actions imitate real events? What limitations does the model have, and do they affect what I can **infer** from the model?

Investigate Self-Assessment	Agree	Not Sure	Disagree
I put the rocks and chalk in the container with a lid.			
I shook the container for several minutes and observed the contents regularly.			
I **inferred** why some of the materials in the container broke down faster than other materials.			

© Harcourt

Name _____

Date _____

Independent Inquiry

First, weigh the chalk that is left after the investigation.
Then, add water to the jar and repeat the test. Compare
the weights of the chalk with and without water. What
conclusion can you draw?

Materials

Here are some materials that you might use.
List additional materials that you need.

- 6 medium-size rocks
- empty plastic juice container with lid (2 qts/1.89 L)
- 2 pieces of chalk

- water

1. What will your investigation test?

_____

_____

2. Record your measurements of the mass of the chalk and the
   mass of the chalk with water.

	Mass (g)
Chalk without water	
Chalk with water	

3. What can you conclude about how water affects the chalk?

_____

_____

Name _____

Date _____

Testing Soil

Materials

measuring
scoop

sand

2 large jars with
wide mouths and lids

potting
soil

250 mL
measuring
cup

water

Procedure

1. Place several scoops of sand in a jar. Place an equal amount of potting soil in another jar.

2. Put 200 mL of water in the measuring cup.

3. Slowly pour the water into the sand. Stop when water starts to puddle on top.

4. Record how much water you used.

5. Repeat Steps 2, 3, and 4 for the potting soil.

Material	Water Used
Sand	
Potting Soil	

© Harcourt

Draw Conclusions

1. **Compare** the amounts of water the two types of soil absorbed. **Infer** where the water you poured into the soil went.

   _____

   _____

   _____

2. **Inquiry Skill—*Compare*** When scientists do an experiment, they often do it in more than one way and **compare** results. What do your results tell you about how the size of particles and spaces between them compare in sand and potting soil?

   _____

   _____

   _____

> **Inquiry Skill Tip**
>
> When you make a calculation from data or measurements, do the calculation twice and **compare** the answers. If the two answers do not agree, do the calculation a third time to confirm the right answer.

Investigate Self-Assessment	Agree	Not Sure	Disagree
I set up the jars with soil or sand, and then added water according to the directions.			
I used the measuring cup to measure the water I used each time.			
I **compared** the amounts of water I used with the sand and the soil.			

© Harcourt

Independent Inquiry

Repeat the investigation, using a different type of soil.
Predict how the new results will compare with the results of
the Investigate. Was your prediction correct?

Materials

Here are some materials that you might use.
List additional materials that you need.

- measuring scoop
- large jar with wide mouth and lid
- 250 mL graduate
- water

1. Predict how the new soil sample will compare with the sand and
   the first soil sample from the Investigate.

   _____

   _____

2. Record your measurements in the table below.

Soil		
Water in Graduate (mL) originally	Water left in Graduate (mL)	Difference
200		

3. Summarize the results of your investigation. How did the new
   soil sample compare to the sand and the original soil sample?

   _____

   _____

© Harcourt

Name _____

Date _____

Make a Landform Model

Materials

pencil   modeling clay   heavy cardboard

Procedure

1 Look for a landform in your area. It might be a mountain, hill,
   dune, valley, plateau, canyon, or cliff.

2 **Observe** the landform's shape and size. Sketch the landform in
   the space below.

3 Get a piece of modeling clay from your teacher. Place it on
   the cardboard.

4 Use clay and your sketch of the landform to **make a model**.

Sketch of Landform	Predictions

© Harcourt

Name _____

Draw Conclusions

1. Which type of landform did you **make a model** of with the clay?

   _____

2. **Predict** how the landform might change in the future. What might cause the change?

   _____

   _____

> **Inquiry Skill Tip**
>
> As you build your model, pay attention to the way it looks. **Observe** the changes that occur as you build the model that resembles a local landform.

3. **Inquiry Skill—*Observe*** Scientists often **observe** objects in nature and then use models to understand them better. How did **observing** the model help you understand the landform you chose?

   _____

   _____

   _____

Investigate Self-Assessment	Agree	Not Sure	Disagree
I drew and constructed the landform according to the directions for this investigation.			
I washed my hands after working with the clay.			
I used the picture of the landform to help me **make a model**.			

Name _____

Date _____

Independent Inquiry

Use the information on a topographic map to make a model
of one of the landforms shown on the map.

Materials

Here are some materials that you might use.
List additional materials that you need.

- pencil
- heavy cardboard
- modeling clay
- topographic map

1. Describe what your model will show.

_____

_____

_____

2. Use this space to copy the topographic map and to draw your
model. Label the high and low places on the map and model.

Topographic Map	Drawing of Your Model

3. Could you construct a topographic map based on a model?
Explain how you would do it.

_____

_____

_____

© Harcourt

Name _____

Date _____

Volcanic Eruptions

Materials

2-liter plastic bottle   small piece of modeling clay   aluminum pie plate

funnel   puffed rice cereal   air pump

Procedure

1 Ask your teacher to make a hole near the bottom of a bottle. Use clay to stick the bottom of the bottle to a pie plate.

2 Use a funnel to add cereal to the bottle until it is 1/4 full.

3 Attach an air pump to the hole in the bottle. Put a piece of clay around the hole to seal it.

4 Pump air into the bottle. **Observe** what happens.

Action	Observation

© Harcourt

Draw Conclusions

1. What happened to the cereal when you pumped air into the bottle?

   _____

   _____

2. **Predict** how you could model a very large eruption.

   _____

   _____

3. Inquiry Skill—*Use models* Scientists often **use models** to help them understand things that happen in nature. How does the bottle model an erupting volcano?

   _____

   _____

   _____

Sometimes when you **make a model**, what you really want to see is how the model behaves. Making the model look realistic isn't as important. When you plan a model, begin by figuring out what you want to learn from it.

Investigate Self-Assessment	Agree	Not Sure	Disagree
I built and tested the model according to the directions for this investigation.			
I was careful to use the air pump in a safe way.			
I **made a model** to help me understand how volcanoes work.			

Name _____

Date _____

Independent Inquiry

Make models using fine sand and gravel to test the
hypothesis: A volcano that forms from thick lava is steeper.

Materials

Here are some materials that you might use.
List additional materials that you need.

- 2-liter plastic bottle
- small piece of modeling clay
- aluminum pie plate
- funnel
- air pump
- puffed rice cereal
- fine sand
- gravel

1. Describe how you will test the hypothesis that a volcano will be
   steeper if it is formed by thicker lava.

   _____

   _____

2. Use this space to make a drawing of your volcano model.

3. How does this model make learning about volcanoes easier?

   _____

   _____

   _____

© Harcourt

Name _____

Date _____

Sets of Animal Tracks

Materials

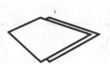
poster
board

markers, crayons,
and colored pencils

animal footprint
stamps

inkpad

Procedure

1 On the poster board, draw a picture of an area where you might find animal tracks, such as a riverbank or a sandy beach.

2 Each person in your group should choose a different animal. Using an inkpad and stamps or other materials, mark the animal's tracks on the poster board. Keep a record of which animal made tracks first, second, third, and so on.

3 Trade finished poster boards with another group. Figure out the order in which the other group's tracks were made. **Record** your conclusions in an ordered list. Give reasons for the order you choose. **Compare** your conclusions with those of other groups.

Order in Which the Other Group's Tracks were Made	Reasons for the Order You Chose

Draw Conclusions

1. Did all the animals move in the same way? How could you tell what kind of animal made the tracks?

_____

_____

_____

2. **Inquiry Skill—*Predict*** Scientists often examine an ecosystem at different times of day to see different animals. **Predict** the different animal tracks you might see if your picture was at night.

_____

_____

_____

Inquiry Skill Tip

Sometimes it is difficult or impossible to observe objects or events directly. Drawing pictures based on the information you have available can help you **predict** the way things behave and work together. Discussing your observations helps you learn because you can compare what you know with things that others know.

Investigate Self-Assessment	Agree	Not Sure	Disagree
I followed the directions for this investigation.			
I used stamps to make animal footprints.			
I made an ordered list of the tracks based on my **observations**.			

© Harcourt

Name _____

Date _____

Independent Inquiry

Make tracks on a sheet of paper. Have a classmate infer
from the tracks how the animal moves. Does it slither, walk,
or jump?

Materials

Here are some materials that you might use.
List additional materials that you need.

- paper
- markers, crayons, and colored pencils
- animal footprint stamps
- inkpad

1. What is the purpose of your investigation?

_____

2. Show your animal tracks in the box below, and write whether it
   slithers, walks, or jumps.

3. Based on your observations of animal tracks, what conclusions can
   you draw about differences in the appearance of animal tracks?

_____

_____

© Harcourt

Name _____

Date _____

Build a Model Solar System

1. Observe and Ask Questions

The solar system contains many objects that revolve around other objects. Planets revolve around the sun. Moons revolve around planets. Not all planets are the same distance from the sun. List questions about how objects revolve in the solar system. Then circle a question you want to investigate.

_____

2. Form a Hypothesis

Write a hypothesis. A hypothesis is a scientific explanation that you can test.

_____

3. Plan an Experiment

Identify and Control Variables To plan your experiment, you must first identify the important variables. Complete the statements below.

The variable I will change is

_____

The variable I will observe or measure is

_____

The variables I will keep the same, or *control*, are

_____

© Harcourt

Name _____

Develop a Procedure and Gather Materials Write the steps you will follow to set up an experiment and collect data.

_____

_____

_____

_____

_____

_____

_____

_____

Use extra sheets of blank paper if you need to write down more steps.

Materials List Look carefully at all the steps of your procedure, and list all the materials you will use. Be sure that your teacher approves your plan and your materials list before you begin.

_____

_____

© Harcourt

4. Conduct the Experiment

Gather and Record Data Make observations to collect data.
Make copies of the chart below or a chart you design to
record your data. **Observe** carefully. **Record** all of your
observations. Also note anything unusual or unexpected.

Name of Planet	Length of String (cm)	Time Required to Revolve (sec)

Other Observations

_____

_____

_____

© Harcourt

Interpret Data Study your data. Make a graph that shows the relationship between the distance a planet is from the sun and how long it takes the planet to revolve around the sun.

5. **Draw Conclusions and Communicate Results**

   Compare the hypothesis with your observations, and then answer these questions.

   1. Was your hypothesis supported by the results of the experiment? Explain.

   _____

   2. How would you revise the hypothesis? Explain.

   _____

   _____

   3. What else did you **observe** during the experiment?

   _____

   _____

Prepare a presentation for your classmates to **communicate** what you have learned. Display your data table.

Investigate Further

Write another hypothesis that you might investigate.

_____

_____

© Harcourt

Name _____

Date _____

From Salt Water to Fresh Water

Materials

500 mL of warm water

salt

spoon

cotton swabs

large clear bowl

small glass jar   plastic wrap   large rubber band   small ball   masking tape

Procedure

1 Stir two spoonfuls of salt into the warm water. Dip a cotton swab into the mixture. Touch the swab to your tongue. **Record** what you **observe. CAUTION: Do not share swabs. Throw the swab away.**

2 Put the jar in the center of the large bowl. Pour the salt water into the bowl. Be careful not to get any salt water in the jar.

3 Put plastic wrap over the bowl. The wrap should not touch the jar. Use the rubber band to hold the wrap in place.

4 Put the ball on the wrap over the jar. Make sure the wrap doesn't touch the jar.

5 Mark the level of the salt water with a piece of tape on the outside of the bowl. Put the bowl in a sunny spot for one day.

6 Remove the wrap and ball. Use a clean swab to taste the water in the jar and bowl. **Record** what you **observe.**

Liquid	Taste
Water with salt	
Water in jar	
Water in bowl	

© Harcourt

Draw Conclusions

1. What did you **observe** during the investigation?

   _____

   _____

   _____

2. **Inquiry Skill—***Infer* Scientists **infer** based on what they **observe**. What can you **infer** is a source of fresh water for Earth?

   _____

   _____

Inquiry Skill Tip

The types of things you **infer** depend on the kind of investigation you have. Sometimes you have to look for a cause and effect. Consider the things you **observed** in the investigation and try to understand what caused them.

Investigate Self-Assessment	Agree	Not Sure	Disagree
I followed the directions for observing the taste of the water.			
I did not share the cotton swabs, and I threw them away after I used them.			
I was able to **infer** the source of Earth's fresh water.			

Independent Inquiry

What would happen if you left the bowl and jar in the sun for several days? Write a hypothesis. Try it!

Materials

- container of warm water
- salt
- spoon
- cotton swabs
- large clear bowl
- small glass jar
- plastic wrap
- large rubber band
- small ball

1. Write your hypothesis.

_____

_____

_____

2. What did you observe in your investigation?

_____

_____

_____

3. Did the results of the investigation support your hypothesis? Explain.

_____

_____

Name _____

Date _____

Modeling a Flood

Materials

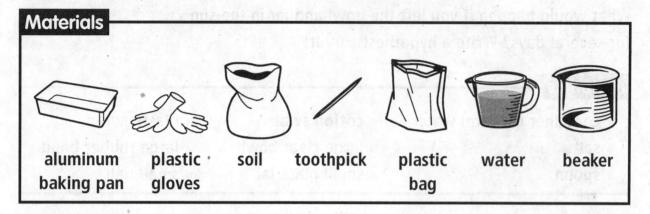

| aluminum baking pan | plastic gloves | soil | toothpick | plastic bag | water | beaker |

Procedure

1. Put on gloves and fill the aluminum baking pan halfway with soil. Make a path in the soil to form a "river channel" that runs through the center of the pan. Build up some small hills around the river channel. Press the soil in place.

2. Use the toothpick to poke several holes in the bottom of the plastic bag.

3. **Measure** 150 mL of water in the beaker. One partner should hold the plastic bag over the pan while you slowly pour the water into the bag. Let the water drip over the pan to **model** a rainy day. **Record** what you **observe**.

4. Repeat Step 3 several times until the pan becomes three-fourths full of water.

Rainy Day	Observations
1	
2	
3	
4	

© Harcourt

Draw Conclusions

1. What happened to the soil in the pan after the first "rainy day"? What happened after the last "rainy day"?

   _____

   _____

   _____

   _____

   _____

2. Inquiry Skill—*Gather/Record/Interpret Data*
   Scientists often **gather, record,** and **interpret data** to understand how things work. **Interpret** what you observed and recorded using your model. What do you think causes floods?

   _____

   _____

Investigate Self-Assessment	Agree	Not Sure	Disagree
I followed the directions for this investigation.			
I **observed** what happened to the soil on "rainy days."			
I **gathered and interpreted data** to learn about one way floods occur.			

© Harcourt

Name _____

Date _____

Independent Inquiry

Would the results be the same if there were several days between each rainfall? Plan and conduct a simple investigation to find out.

Materials

- aluminum baking pan
- soil
- toothpick
- plastic bag
- water
- beaker

1. Write your hypothesis.

_____

_____

2. Use the table below to record your observations.

Rainy Day	Observations
1	
2	
3	
4	

3. Did the results of the investigation support your hypothesis? Explain.

_____

_____

_____

© Harcourt

Name _____

Date _____

Heating Land and Water

Materials

2 small plastic or foam cups | 2 thermometers | dark soil or sand | water | stopwatch | light source with 100-W bulb or greater

Procedure

1 Fill one cup with dark soil or sand. Fill the second cup with water. Place a thermometer upright in each cup.

2 Time one minute using the stopwatch. Then **measure** and **record** the temperatures of the two cups.

3 Place the cups under the light. Make sure that both cups get an equal amount of light.

4 After the cups have been under the light for five minutes, **measure** and **record** their temperatures. Repeat this step every 5 minutes for the next 15 minutes.

5 Turn the lamp off. Time 5 minutes, and then **measure** and **record** the temperatures of the cups. Repeat this step every 5 minutes for the next 10 minutes.

© Harcourt

	Lamp On Time (min)					Lamp Off Time (min)		
Contents of Cup	0	5	10	15	20	5	10	15
Soil or sand								
Water								

Name _____

Draw Conclusions

1. Describe how the soil and water heated differently. How did they cool differently?

   _____

   _____

2. **Inquiry Skill—***Hypothesize* Scientists use what they **observe** to form a **hypothesis**. Use your **observations** from this investigation to **hypothesize** how the weather on Earth would be different if Earth's surface was mostly land instead of water.

   _____

   _____

   _____

Inquiry Skill Tip

A **hypothesis** should always be an idea that can be tested. When you **hypothesize** about Earth's processes, the ideas could be tested using a model.

Investigate Self-Assessment	Agree	Not Sure	Disagree
I followed the directions for modeling the heating of land and water.			
I used the thermometers to measure the temperatures of the cup contents.			
I **hypothesized** how the weather on Earth would be different if Earth's surface was mostly land.			

Name _____

Date _____

Independent Inquiry

Does wet soil heat differently from dry soil? Conduct an experiment to find out.

Materials

Here are some materials that you might use.
List additional materials that you need.

- 2 small plastic or foam cups
- 2 thermometers
- dark soil or sand
- water
- stopwatch
- light source with 100-W bulb or greater

1. What will you learn from the investigation?

_____

_____

2. Draw a data table for your investigation in the space below. Record your data in the table.

3. What conclusions can you draw from the investigation?

_____

_____

_____

© Harcourt

Name _____

Date _____

Making a Barometer

Materials

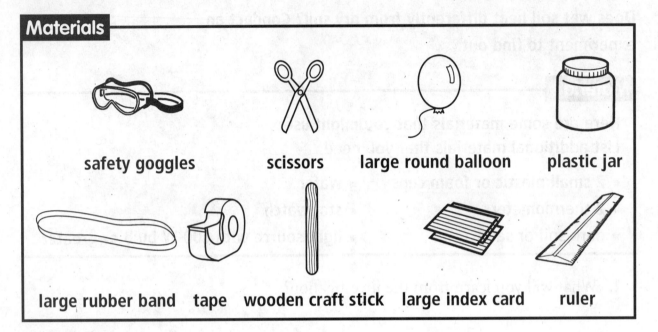

safety goggles   scissors   large round balloon   plastic jar

large rubber band   tape   wooden craft stick   large index card   ruler

Procedure

1. **CAUTION: Wear safety goggles.** Be careful when using scissors. Use the scissors to cut the neck off the balloon.

2. Have your partner hold the jar while you stretch the balloon over the open end. Secure the balloon with the rubber band.

3. Tape the craft stick to the top of the balloon. More than half of the craft stick should extend beyond the jar's edge.

4. On the blank side of an index card, draw a line and label it *Day 1*. Tape the card to a wall. The line should be at the same height as the stick on your barometer. Next to it, **record** the current weather.

5. Air pressure is the force of air pressing down on Earth. **Measure** air pressure by marking the position of the wooden stick on the index card for the next four days. Label the marks *Days 2–5*. **Record** the pressure and weather each day on the chart.

Day	Pressure	Weather
1		
2		

© Harcourt

Draw Conclusions

1. How did the air pressure change? What might cause changes in air pressure?

_____

_____

_____

2. **Inquiry Skill—***Measure* Scientists use instruments to **measure** weather data. **Infer** how your barometer works.

_____

_____

_____

_____

_____

Inquiry Skill Tip

The instruments you use to make a measurement depend on what you need to **measure**. Sometimes you need very precise instruments. Other times, less precise instruments are acceptable. The barometer used in this investigation isn't precise, but it is useful for understanding weather forecasting.

Investigate Self-Assessment	Agree	Not Sure	Disagree
I followed the directions for making a barometer.			
I wore goggles and obeyed safety rules when using the scissors.			
I **measured** air pressure using the barometer.			

© Harcourt

Name _____

Date _____

Independent Inquiry

Track changes in air pressure and weather for five more
days. What can you infer is the relationship between air
pressure and type of weather?

Materials

- safety goggles
- scissors
- large round balloon
- ruler
- large rubber band
- tape
- wooden craft stick
- large index card
- plastic jar

1. What is the purpose of your investigation?

_____

_____

2. Use the table below to **record** your observations.

Day	Air Pressure	Weather Observations
1		
2		
3		
4		
5		

3. From the data and observations, what can you infer is the
relationship between air pressure and type of weather?

_____

_____

Seasons and Sunlight

Materials

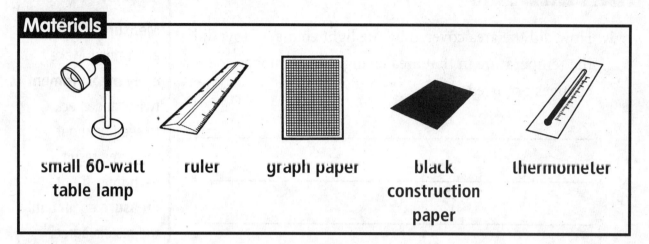

small 60-watt table lamp   ruler   graph paper   black construction paper   thermometer

Procedure

1 Work with a partner. Shine the lamp straight down from a height of 30 cm onto a piece of graph paper. Draw an outline of the lit area. Label it Step 1.

2 Repeat Step 1, this time placing the lamp at an angle. Label this outline Step 2.

3 Now use the black paper instead of graph paper. Shine the lamp straight onto it. After 15 minutes, **measure** the temperature of the paper in the lit area. **Record** the temperature on the sheet of graph paper labeled Step 1.

4 Now angle the light as you did in Step 2. Again, **measure** the temperature of the lit area after 15 minutes. **Record** the temperature on the sheet of graph paper labeled Step 2.

Step	Lamp Angle	Temperature
1	straight down	
2	tilted	

© Harcourt

Name _____

Draw Conclusions

1. How did the area covered by the light change? How did the temperature in that area change? Explain why these changes occurred.

_____

_____

_____

_____

2. Inquiry Skill—*Measure* How could scientists **measure** the effect the sun has on seasons?

_____

_____

_____

Investigate Self-Assessment	Agree	Not Sure	Disagree
I followed the directions for this investigation.			
I used the ruler to **measure** the height of the lamp.			
I **measured** the temperature of the lit area during the investigation.			

> **Inquiry Skill Tip**
>
> **Measure** accurately! Make every **measurement** twice, and check to see if you get the same result each time. If not, **measure** again until the results agree.

© Harcourt

Independent Inquiry

Try your experiment on the real thing! Pick a sunny
spot outdoors, and measure and record its temperature
throughout the day. Why do you think it changes?

Materials

- thermometer

1. Predict how the temperature will change during the day.

   _____

   _____

   _____

2. Make a table to record temperature data and the time of day.

3. Summarize the results of your investigation. Why do you think
   the temperature changes?

   _____

   _____

   _____

© Harcourt

Name _____

Date _____

Distances Between Planets

Materials

4-m length of string   9 different-colored markers   tape measure

Procedure

1 At one end of the string, make a large knot. This knot will stand for the sun as you **make your model.**

2 Earth averages a distance of 1 AU (astronomical unit) from the sun. In your model, 1 AU will equal 10 centimeters. Use the tape measure to **measure** Earth's distance from the knot that represents the sun. Use a marker to mark this point on the string. **Record** on the table which color you used.

3 Complete the Scale Distance column of the table. Repeat Step 3 for each of the planets. Use a different color for each planet.

Planet	Average Distance from Sun(km)	Average Distance from Sun(AU)	Scale Distance (cm)	Planet's Diameter (km)
Mercury	58 million	4/10	4	4,876
Venus	108 million	7/10	7	12,104
Earth	105 million	1		12,756
Mars	228 million	2		6,794
Jupiter	778 million	5		142,984
Saturn	1429 million	10		120,536
Uranus	2871 million	19		51,118
Neptune	4500 million	30		49,532
Pluto*	5900 million	39		2,274

*In 2006, scientists classified Pluto as a "dwarf planet."

Draw Conclusions

Inquiry Skill Tip

Using numbers helps you make precise comparisons between a model and the real thing. Numbers also can help you see patterns, for example, by putting data in order by size.

1. In your **model**, how far away from the sun is Mercury? How far away is Pluto?

_____

2. Why do scientists use AUs to **measure** distances in the solar system?

_____

_____

_____

3. **Inquiry Skill—**_Use Numbers_ How does it help to **use numbers** instead of using real distances?

_____

_____

Investigate Self-Assessment	Agree	Not Sure	Disagree
I followed the directions for making a model of distances between planets.			
I used the tape measure to measure distances on the string.			
I **used numbers** showing the relative distance of each planet from the sun.			

© Harcourt

Name _____

Date _____

Independent Inquiry

Use a calculator to make models of planet diameters. Use
1 cm as Earth's diameter. Then divide the diameters of other
planets by Earth's. Make a scale drawing.

Materials

Here are some materials that you might use.
List additional materials that you need.

- 9 different-colored markers
- tape measure
- paper

1. Choose any planet. Explain how you will calculate its diameter
   on the scale drawing.

   _____

   _____

   _____

2. Make a table to record the diameter you will use for each planet
   in the scale drawing.

3. Summarize what you learned from the model about the
   diameters of the planets.

   _____

   _____

   _____

© Harcourt

Shining Constellations

Materials

- black construction paper
- reference book of constellations
- sharp pencil

Procedure

1. Work with your partner to find and research a constellation

2. Use the pencil to mark the constellation on your sheet of black construction paper.

3. Use the sharp point of your pencil to poke a small hole where you placed each dot for a star.

4. Ask your teacher to place your sheet of construction paper with your constellation on an overhead projector. See if your classmates can identify the constellation.

5. Explain the story that is related to your constellation. Discuss the constellations that your classmates chose to present.

Constellation	Story

© Harcourt

Name _____

Draw Conclusions

Inquiry Skill Tip

Think about the question you wanted to answer with the investigation. That will help guide you when you **plan an investigation**.

1. What did you observe when your piece of construction paper was placed on the overhead projector?

   _____

   _____

2. **Inquiry Skills**—Navigators used constellations to help them guide ships before there were accurate tools to do so. They **observed** the movement of the stars in the sky. How would you **plan and conduct a simple investigation** to find out if the appearance of the stars changed with your location?

   _____

   _____

   _____

   _____

Investigate Self-Assessment	Agree	Don't Know	Disagree
I identified and researched a constellation			
I followed instructions to make a model of my constellation.			
I accurately sketched the stars.			
I worked well with my partner.			

© Harcourt

Name _____

Date _____

Independent Inquiry

**Find your constellation in the night sky. Observe the
movement of the constellation through the sky for a month.**

Materials

- **Paper**
- **Pencil**

1. Observe the nighttime sky. Sketch your observations.

2. At the same time each night for four weeks, observe the nighttime
   sky, and sketch your observations.

3. **Compare** your results. What can you **conclude** about the
   movement of stars in the night sky?

_____

_____

_____

_____

Name _____

Date _____

Color and Light Energy Absorption

1. Observe and Ask Questions

How does color affect the amount of light energy an object will absorb? List questions you have about color and light energy absorption. Circle a question to investigate.

_____

_____

2. Form a Hypothesis

Write a hypothesis. A hypothesis is a scientific explanation you can test.

_____

3. Plan an Experiment

To plan your experiment, you must first identify the important variables. Complete the statements below.

Identify and Control Variables

The variable I will change is _____

The variables I will observe or measure are _____

The variables I will keep the same, or *control*, are _____

_____

Develop a Procedure and Gather Materials Write the steps
you will follow to set up an experiment and collect data.

_____

_____

_____

_____

_____

_____

_____

_____

_____

_____

_____

Use extra sheets of blank paper if you need to write down more steps.

Materials List Look carefully at all the steps of your procedure,
and list all the materials you will use. Be sure that your teacher
approves your plan and your materials list before you begin.

_____

_____

_____

Name ______________________

Inquiry Log

4. Conduct the Experiment

Gather and Record Data Follow your plan and collect data. Use the chart below or a chart you design to record your data. **Observe** carefully. **Record** your observations, and be sure to note anything unusual or unexpected.

Location: ______ Type of Material: ______ Light Source: ______												
Color of Material:	Temperature (°C every 5 minutes for 60 minutes)											
	5	10	15	20	25	30	35	40	45	50	55	60
Control												

Other Observations

____

____

____

____

Interpret Data Make a graph of the data you have collected. Plot the graph on a sheet of graph paper, or use a software program.

5. **Draw Conclusions and Communicate Results**

Compare the **hypothesis** with the data and the graph, and then answer these questions.

1. Given the results of the experiment, do you think the hypothesis is true? Why or why not? Explain.

_____

_____

2. How would you revise the hypothesis? Explain.

_____

_____

3. What else did you **observe** during the experiment?

_____

_____

Prepare a presentation for your classmates to **communicate** what you have learned. Display your data tables and graphs.

Investigate Further

Write another hypothesis that you might investigate.

_____

_____

© Harcourt

Name _____

Date _____

Measuring the Densities of Liquids

Materials

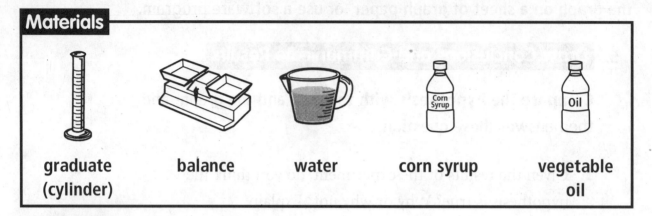

graduate (cylinder)   balance   water   corn syrup   vegetable oil

Procedure

1 Make sure the cylinder is empty, clean, and dry. Then use the balance to find its mass. **Record** the mass.

2 Add 10 mL of water to the cylinder. **Measure** and **record** the mass of the graduate cylinder. Empty and dry the cylinder again.

3 Repeat Step 2, using 10 mL of vegetable oil.

4 Repeat Step 2, using 10 mL of corn syrup.

5 Subtract the mass of the empty cylinder from each of the masses you **measured** in Step 2. **Record** each result.

6 To find the densities, divide the mass of each liquid by its volume, 10 mL. **Record** and **compare** the densities.

Liquid	Mass (cylinder only) (g)	Mass (cylinder and liquid) (g)	Mass (liquid only) (g)	Volume (mL)	Density (g/mL)
Water				10	
Corn syrup				10	
Vegetable oil				10	

© Harcourt

Draw Conclusions

1. Which liquid has the greatest density? Which has the least? Compare the amount of matter in each liquid.

   _____

   _____

   _____

   _____

   _____

2. **Inquiry Skill—*Display Data*** Display data by making a bar graph that shows the density of each liquid you measured.

   _____

   _____

   _____

Investigate Self-Assessment	Agree	Not Sure	Disagree
I followed the directions for this investigation.			
I used the graduate and the balance to measure the mass and volume of each liquid.			
I **displayed data** by making a bar graph of the densities.			

Name _____

Date _____

Independent Inquiry

Tint 10 mL of water red. Pour it into the cylinder. Then add
10 mL of corn syrup. Observe what happens. How do you
explain your observation?

Materials

- graduate
- water
- red food coloring
- corn syrup

1. Predict what will happen when you put 10 mL of red water and
   10 mL of corn syrup in the same graduated cylinder.

   _____

   _____

2. Describe what you observed during the investigation.

   _____

   _____

   _____

3. Use what you have learned about density to explain the results of
   this investigation.

   _____

   _____

   _____

© Harcourt

Melt, Boil, Evaporate

Materials

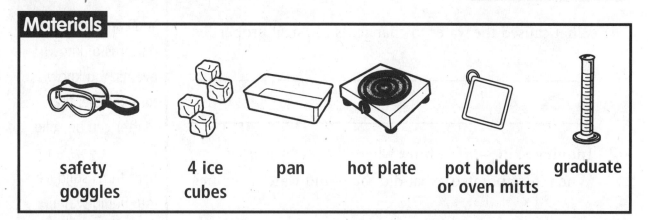

| safety goggles | 4 ice cubes | pan | hot plate | pot holders or oven mitts | graduate |

Procedure

1 Draw the ice cubes. Describe their physical properties, including how they look and feel.

2 **Caution: Put on safety goggles.** Put the ice in the pan, and carefully heat the pan on the hot plate. If you must touch the pan, use a pot holder. **Predict** the changes you expect to see in the ice cubes.

3 When the ice cubes melt, pour the water into the cylinder. **Record** its volume. Use pot holders as you pour.

4 Pour the water back into the pan. Put it on the hot plate again, and let the water boil. **Predict** what will happen this time.

Caution: Remember to turn off the hot plate. Remove the pan from the heat before it is dry. Place it on a burn-proof surface.

Description:	
Prediction:	
Volume:	
Prediction:	

© Harcourt

Name _____

Draw Conclusions

1. What caused the water to change its physical properties?

   _____

   _____

2. **Inquiry Skill—*Infer* Infer** where the water is now.
   When it evaporated, what did the liquid water become?

   _____

   _____

Inquiry Skill Tip

You can **infer** from patterns of known events. You know, for example, that matter can be solid, liquid, or gas. Since heat turns solid ice into liquid water, you can infer that it turns liquid water into a gas.

Investigate Self-Assessment	Agree	Not Sure	Disagree
I followed the directions for this investigation.			
I followed safety rules for handling hot objects.			
I **inferred** the location of the water after it evaporated.			

Independent Inquiry

Using oven mitts, pick up an ice cube in each hand. Push the cubes together while still wearing the oven mitts. Explain what you observe.

Materials

- **2 ice cubes**
- **pot holders or oven mitts**

1. Predict what will happen when you push the ice cubes together.

_____

_____

2. Record your observations of putting the two ice cubes together.

_____

_____

3. Explain your observations.

_____

_____

_____

Which Solids Will Dissolve?

Materials

water   teaspoon   sand   4 clear containers

salt   sugar   baking soda

Procedure

1 Half-fill each container with water.

2 Put 1 spoonful of sand into one container. **Observe** and **record** what happens.

3 Stir the mixture for 1 minute, and then **record** what you see.

4 Repeat Steps 2 and 3, using salt, sugar, and baking soda. **Observe** and **record** all the results.

Material	Before Stirring	After Stirring
Sand		
Salt		
Sugar		
Baking soda		

© Harcourt

Draw Conclusions

1. Which solid dissolved the most? Which did not dissolve at all?

_____

_____

_____

_____

2. Inquiry Skill—*Plan and Conduct a Simple Investigation* Scientists often **plan a simple investigation** to test an idea quickly. What idea did this activity test? What is another simple investigation you could do with these materials?

_____

_____

Inquiry Skill Tip

A **simple investigation** should use only a few materials. You should be able to finish it in 20–30 minutes. Your **plan** should list all the materials you need, the steps you will follow, and what you will measure or observe.

Investigate Self-Assessment	Agree	Not Sure	Disagree
I observed each of the four materials in water.			
I used the teaspoon to measure the amount of each material.			
I **compared** how the four materials dissolved in water.			

© Harcourt

Name _____

Date _____

Independent Inquiry

Dissolve table sugar and powdered sugar in separate
containers of water. Compare how quickly the two kinds of
sugar dissolve. Explain your results.

Materials

- water
- teaspoon
- 2 clear containers
- stirrer
- table sugar
- powdered sugar

1. What will your investigation test?

_____

_____

2. Use the table below to record your observations.

Material	Observations
Table sugar	
Powdered sugar	

3. Compare how quickly the two kinds of sugar dissolve. Explain
   your results.

_____

_____

© Harcourt

Name _____

Date _____

A Solution to the Problem

Materials

iodized salt	kosher salt	sea salt	granulated sugar	powdered sugar
brown sugar	6 spoons	6 plates	6 plastic cups	water

Procedure

1 Place a small amount of each kind of salt and each kind of sugar on its own plate.

2 **Compare and contrast** each sample for color and texture. **Record** your **observations**.

3 **Compare and contrast** the grain size of each sample. **Record** your **observations**.

4 Place the same amount of water in each of six cups. Use a clean spoon to place the same amount of each sample in its own cup. Stir. Use a table like this to **record** your **observations**.

Sample	Color	Texture	Grain Size	Reaction in Water
Iodized salt				

© Harcourt

Draw Conclusions

Inquiry Skill Tip

When one factor or event has no effect on the other, you must **draw the conclusion** that there is no relationship between them.

1. Which samples—the light-colored ones or the darker-colored ones—mixed into the water more quickly?

   _____

   _____

2. Which samples—the ones with larger grains or the ones with smaller grains—mixed into water more quickly?

   _____

   _____

3. **Inquiry Skill—*Draw Conclusions*** Scientists interpret data to **draw conclusions**. What can you conclude about how color and grain size affect the speed with which a sample mixed into water?

   _____

   _____

Investigate Self-Assessment	Agree	Not Sure	Disagree
I observed the color, texture, and grain size of each type of salt and sugar.			
I followed safety rules by not tasting any of the materials used in this investigation.			
I **drew a conclusion** about how color and grain size affected the speed in which a sample mixed into water.			

Name _____

Date _____

Independent Inquiry

Sequence the samples by how much of them mixed into water. Predict where sugar cubes will fit on your list. Test your prediction.

Materials

- spoon
- plastic cup
- water
- sugar cube

1. Make a table to sequence the salt and sugar samples by how much of them mixed into water.

2. Predict where a sugar cube will fit in the table. Explain your answer.

   _____

   _____

   _____

   _____

3. Summarize the results of your investigation. How do they compare to your prediction?

   _____

   _____

   _____

Name _____

Date _____

Drop by Drop

Materials

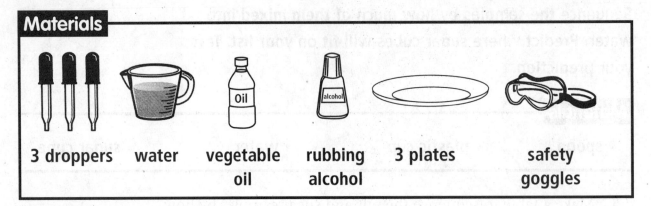

| 3 droppers | water | vegetable oil | rubbing alcohol | 3 plates | safety goggles |

Procedure

1 CAUTION: **Wear safety goggles**. Place 3 drops of water on one plate, 3 drops of vegetable oil on the second plate, and 3 drops of rubbing alcohol on the third plate. Be sure to use a different dropper for each liquid.

2 **Record** your **observations** of each liquid.

3 Repeat Step 2 every half hour for the rest of the school day.

Observation	Water	Oil	Alcohol
1			
2			
3			
4			
5			
6			
7			
8			
9			

© Harcourt

Draw Conclusions

1. What did you **observe** at the end of the day?

_____

_____

_____

2. Inquiry Skill—*Hypothesize* When scientists give a possible explanation for what they observe, they are making a **hypothesis**. Then scientists test the hypothesis. What **hypothesis** can you make from your observations?

_____

_____

_____

> **Inquiry Skill Tip**
>
> When you **hypothesize**, you should use facts and logic to explain *all* of the observations. Your **hypothesis** from the investigation should explain what happened to the water, oil, and alcohol—not just one or two of the liquids.

Investigate Self-Assessment	Agree	Not Sure	Disagree
I observed the three liquids every half hour.			
I wore safety goggles when placing the liquids on the plates.			
I was able to **hypothesize** about the liquids, based on my observations.			

© Harcourt

Name _____

Date _____

Independent Inquiry

**What could you do to test your hypothesis? Plan and carry
out an investigation to find out.**

Materials

- **3 droppers**
- **water**
- **vegetable oil**
- **rubbing alcohol**
- **safety goggles**
- **3 plates**

1. Summarize how you will test your hypothesis.

_____

_____

2. Draw a data table that you can use for your investigation. Record
your observations in the data table.

3. Did your investigation support your hypothesis? If so, explain
how. If not, suggest a new hypothesis.

_____

_____

_____

_____

Name _____

Date _____

Wet Wool

Materials

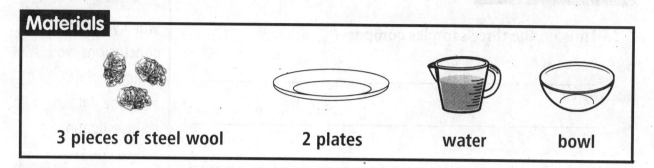

3 pieces of steel wool   2 plates   water   bowl

Procedure

1. Put one piece of steel wool on a plate.

2. Soak another piece of steel wool in water. Then put it on the other plate.

3. Fill the bowl with water, and put the third piece of steel wool in the water. Make sure none of it sticks out above the water.

4. Place all three samples in the same area, away from direct sunlight. Examine them every day for a week. **Record** your **observations**.

Day	Dry Steel Wool	Wet Steel Wool	Steel Wool in Water
1			
2			
3			
4			
5			

Draw Conclusions

1. How do the three samples **compare**?

_____

_____

_____

2. **Inquiry Skill—***Draw Conclusions* Scientists can **draw conclusions** from the results of their experiments. What two things can you conclude caused the changes?

_____

_____

_____

Investigate Self-Assessment	Agree	Not Sure	Disagree
I set up the three samples according to the directions.			
I **observed** the samples every day for a week.			
I **drew a conclusion** about what caused the changes in the steel wool.			

© Harcourt

Investigate
Log

Independent Inquiry

What can you predict will happen if you wet the three samples and place them in direct sunlight? Carry out a test to find out.

Materials

- 3 small pieces of steel wool
- 2 plates
- water
- bowl

1. Write your hypothesis for this investigation.

   _____

   _____

   _____

2. Make a table to use for your investigation. Record the results of your investigation in the table.

3. Did the results of the investigation support your hypothesis? Explain.

   _____

   _____

   _____

© Harcourt

Feel the Vibes

Materials

plastic ruler

Procedure

1 Place the ruler on a desk or tabletop. Let 20 cm of the ruler stick out over the edge of the table.

2 With one hand, press the end of the ruler tightly to the tabletop. With the other hand, flick the other end of the ruler.

3 **Observe** the ruler with your eyes. **Record** your observations.

4 Repeat Step 2. **Observe** the ruler with your ears. **Record** your observations.

5 Change the strength with which you flick the end of the ruler. **Observe** the results and **record** your observations.

6 Change the length of the ruler that hangs over the edge of the tabletop. Repeat Steps 2–5. **Observe** the results and **record** your observations.

Length of Ruler Over Table Edge	How hard the Ruler is Pressed	Observations with Eyes	Observations with Ears
20 cm	Lightly		
20 cm	Stronger		
Greater than 20 cm	Lightly		
Greater than 20 cm	Stronger		

Draw Conclusions

1. What did you observe in Step 3? In Step 4? How do you think these observations are related?

_____

_____

_____

2. Inquiry Skill—*Hypothesize* Hypothesize how changing the ruler affects the sound it makes. Tell how you would test your hypothesis.

_____

_____

_____

Inquiry Skill Tip

Whenever you **hypothesize**, you are making a guide for an investigation. Think about your observations and what conclusions you can draw from them. Based on this, think about other questions you have. Your hypothesis is what you believe is the answer to a question. You can find out if your hypothesis is correct by experimenting.

Investigate Self-Assessment	Agree	Not Sure	Disagree
I followed the directions for this investigation.			
I made observations with my eyes and with my ears.			
I **hypothesized** how changing the ruler affects the sound it makes.			

Name _____

Date _____

Independent Inquiry

Place one ear on the tabletop and cover the other ear with
your hand. Have a partner repeat Steps 1 and 2. What do
you observe?

Materials

- plastic ruler

1. What will your investigation test?

   _____

   _____

2. Describe what you observed when your partner pressed on the
   end of the ruler.

   _____

   _____

   _____

   _____

   _____

3. What can you conclude from your investigation?

   _____

   _____

© Harcourt

Feel the Vibes II

Materials

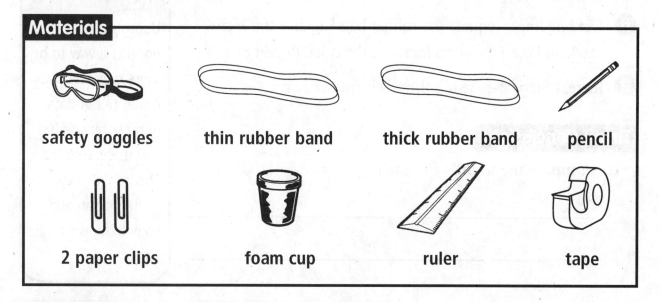

safety goggles   thin rubber band   thick rubber band   pencil

2 paper clips   foam cup   ruler   tape

Procedure

1 CAUTION: **Put on the safety goggles**. Use the pencil to poke a hole in the bottom of the cup.

2 Thread the thin rubber band through the paper clip. Put them in the cup, and then pull the rubber band through the hole.

3 Place the cup on the table upside down. Tape the ruler to the cup as shown in the picture on page 541 of your textbook, with the 1-cm mark at the top. Pull the end of the rubber band over the end of the ruler and tape it to the back.

4 Pull the rubber band to one side, and then let it go. **Observe** the sound. Begin a table like this to **record** your observations.

Rubber Band	Mark on Ruler	Distance Pulled	Observations

© Harcourt

5 Repeat Step 4, but this time, pull the rubber band farther.

6 Use one finger to press the rubber band against the 2-cm mark and then the 4-cm mark on the ruler. Repeat Step 4.

7 Repeat Steps 2–6, using the thick rubber band.

Draw Conclusions

1. **Compare** the sounds you made.

_____

_____

_____

2. **Inquiry Skill**—*Using Numbers* How did **using the numbers** on the ruler help you to give a reason for the sounds that you made?

_____

_____

> **Inquiry Skill Tip**
>
> **Using numbers** is not just a way to be more precise. It is easier to put data in order by size if you **use numbers**. This can give you a clue about why something is happening.

Investigate Self-Assessment	Agree	Not Sure	Disagree
I followed the directions for this investigation.			
I obeyed safety rules and wore safety goggles.			
Based on my observations, I **used numbers** to infer why the different sounds were produced.			

© Harcourt

Name _____

Date _____

Independent Inquiry

Hypothesize how you could use your setup to play a musical scale. Test your hypothesis.

Materials

- safety goggles
- thin rubber band
- thick rubber band
- pencil
- 2 paper clips
- foam cup
- ruler
- tape

1. Write your hypothesis for the investigation.

   _____

   _____

   _____

2. Describe your test and observations.

   _____

   _____

   _____

3. What other way could you make different notes using your setup?

   _____

   _____

© Harcourt

Name _____

Date _____

Do You Hear What I Hear?

Materials

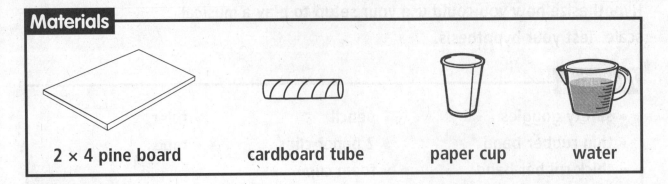

2 × 4 pine board   cardboard tube   paper cup   water

Procedure

1 Have a partner rub a fingernail on a desktop. Listen, and **record** your **observations**. Press one ear against the desktop, and then repeat rubbing the fingernail.

2 Have a classmate hold the board. Press your ear against one end of the board. Then have the classmate lightly rub a fingernail against the wood at the other end. Listen, and **record** your **observations**.

3 Hold one end of the cardboard tube up to your ear. Have a partner lightly rub a fingernail on the other end of the tube. Listen, and **record** your **observations**.

4 Press the side of the paper cup against your ear. Have a partner (continued)

Material	Observations
Desktop through air	
Ear against desktop	
Ear against board	
Ear against cardboard tube	
Ear against empty cup	
Ear against cup with water	

© Harcourt

lightly rub a fingernail on the other side of the cup.
Listen, and **record** your **observations**.

5 Fill the cup with water and repeat Step 4.

Draw Conclusions

1. What differences did you **observe** in the sound
   each time?

   _____

   _____

2. **Inquiry Skill—*Identify Variables*** Which **variable**
   did you change in this procedure? Which **variable** did
   you **observe**?

   _____

   _____

Investigate Self-Assessment	Agree	Not Sure	Disagree
I followed the directions for testing the sounds made through different materials.			
I **observed** the sound in each step of the investigation.			
I **identified variables** that I wanted to observe and the variable that I wanted to change.			

© Harcourt

Name _____

Date _____

Independent Inquiry

Can you observe how sound travels through a liquid other than water? Plan and conduct a simple investigation to find out.

Materials

Here are some materials that you might use.
List additional materials that you need.

- vegetable oil
- water
- paper cup

1. What will your investigation test?

_____

_____

2. Use the data table below to **record** your observations.

Conditions	Observations

3. What can you conclude from your observations?

_____

_____

_____

© Harcourt

How Light Travels

Materials

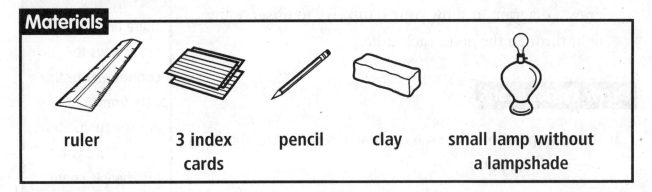

ruler   3 index cards   pencil   clay   small lamp without a lampshade

Procedure

1 Using the ruler, draw lines on each card from corner to corner so that each one has a large X.

2 Use a pencil to make a hole at the center of each X. Stack the cards on top of each other, and use the pencil to make sure the holes are at the same height.

3 Make a clay stand for each card. Stand the cards on the desk, one in front of the other and a few centimeters apart.

4 Place the lamp on the desk, and turn on the light. Look through the holes in the cards. Move the cards until you see the light bulb through all cards at once. Draw a picture to show where the cards are.

Setup 1	Setup 2
Setup 3	**Setup 4**

© Harcourt

5 Move the cards to new places. Each time you move them, draw a diagram to show your setup. Try to **observe** the light through the holes each time.

Draw Conclusions

Inquiry Skill Tip

There are many ways to **communicate** the data from your investigation. Make a list of five ways that people share information.

1. In what position were the cards when you were able to see the light bulb?

_____

_____

2. Inquiry Skills—*Communicate* How did drawing diagrams help you **communicate** your results?

_____

_____

Investigate Self-Assessment	Agree	Not Sure	Disagree
I prepared the cards according to the directions for this investigation.			
I was careful not to look at the light bulb or touch it while it was on.			
I drew diagrams to **communicate** what I learned in this investigation.			

© Harcourt

Independent Inquiry

Would your results be the same if the cards were not on a level surface? Plan and conduct a simple investigation to find out.

Materials

Here are some possible materials. List any additional materials.

- ruler
- 3 index cards
- pencil
- clay
- small lamp without a lampshade

1. Describe how you set up your cards on a surface that isn't level.

_____

_____

2. Use this space to record the arrangements you use.

Setup 1	Setup 2
Setup 3	Setup 4

3. Were your results the same as or different from your investigation using cards on a flat desktop?

_____

_____

© Harcourt

Name _____

Date _____

Build a Thermometer

Materials

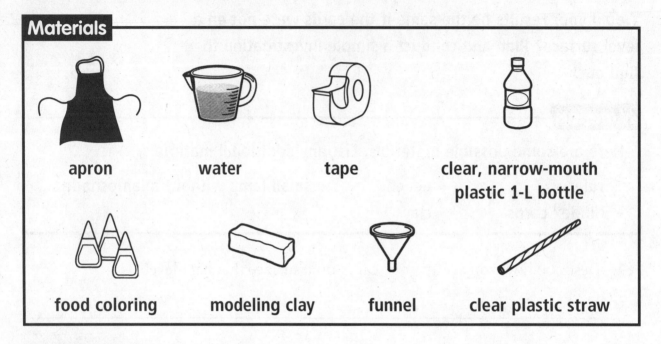

apron	water	tape	clear, narrow-mouth plastic 1-L bottle
food coloring	modeling clay	funnel	clear plastic straw

Procedure

1 Put on an apron. Add water to the bottle until it is about one-third full.

2 Add two or three drops of food coloring. Swirl the bottle to mix it evenly.

3 Put the straw in the bottle. Hold the straw so that the end is in the water and just above the bottom of the bottle. Tape the straw in place at the neck of the bottle.

4 Use clay to seal the top of the bottle so that no air can get in or out. Do not squeeze the bottle.

5 Make a drawing of your thermometer. Show the water level in the straw.

Thermometer originally

Thermometer in a warm place

© Harcourt

6 Put the thermometer in a warm place for five minutes. **Observe** the water level in the straw. Make a drawing to **record** this observation.

Draw Conclusions

Inquiry Skill Tip

To **predict** what will happen under new conditions, think first about what you know about those conditions and what you have observed in the past.

1. Did the water level in the straw change when the thermometer was in a warm place? If so, how did it change?

_____

_____

2. Inquiry Skills—*Predict* Suppose you put your thermometer in a cold place, such as a refrigerator. **Predict** what would happen.

_____

_____

Investigate Self-Assessment	Agree	Not Sure	Disagree
I used the materials to make a thermometer according to the directions for this investigation.			
I obeyed safety rules and did not put any materials into my mouth.			
I predicted how the thermometer would act in a cold place.			

© Harcourt

Independent Inquiry

Predict which places in your classroom will be the warmest
and coolest. Then plan and conduct a simple experiment
to check.

Materials

Here are some possible materials. List any additional materials.

- clear, narrow-mouth plastic 1-L bottle
- clear plastic straw
- water
- funnel

- modeling clay
- food coloring
- tape
- smock

1. Which places do you **predict** will be the warmest and coldest
   places in your classroom?

   _____

   _____

2. Use a table like this to record data for all the places you test.

Location	Height of Water in Thermometer

3. How did the results of your investigation compare to
   your prediction?

   _____

   _____

   _____

Build a Solar Hot Spot

Materials

poster board

scissors

aluminum foil

thermometer

shoe-box lid

clock or watch   hole punch   string   2 sheets of graph paper   glue   ruler

Procedure

1 **CAUTION: Be careful when using scissors.** Cut a piece of poster board 10 cm by 30 cm. Glue foil to one side. Let it dry for at least 10 minutes.

2 Place a thermometer in a shoe-box lid.

3 Place the lid in sunlight. **Record** the temperature each minute for 10 minutes.

4 Punch a hole 2 cm from each end of the poster board. Pull the ends together with string. When the ends are about 20 cm apart, tie the string to hold the shape.

5 Put the curved reflector in the shoe-box lid, with the thermometer in its center. Repeat Step 3.

6 **Make a line graph** of the data from Step 3. Show time on the horizontal axis and temperature on the vertical axis. Then **make a graph** of the data from Step 5.

Time (min)	Temp. (°C)	Time (min)	Temp. (°C)
1		6	
2		7	
3		8	
4		9	
5		10	

© Harcourt

Draw Conclusions

1. Describe the temperature changes shown on each graph.

_____

_____

_____

_____

_____

2. Inquiry Skills—*Interpret Data* Interpret the data.
What do you infer caused the differences in the
temperatures on the two graphs?

_____

_____

Inquiry Skill Tip

Look only at the data you have collected when you are **interpreting the data**. Don't base your results on what you expect or what other students may have measured.

Investigate Self-Assessment	Agree	Not Sure	Disagree
I used the materials to build and test a reflector according to the directions.			
I used the scissors carefully.			
I **inferred** what caused the difference between the temperatures.			

© Harcourt

Name _____

Date _____

Independent Inquiry

Repeat the experiment using black construction paper instead of foil. Make a chart to compare your results. Explain the difference in your results from the two surfaces.

Materials

Here are some possible materials. List any additional materials.

- black construction paper
- 2 sheets of graph paper
- poster board
- shoe-box lid

- thermometer
- scissors
- hole punch

- clock or watch
- glue
- string

1. Describe what you are comparing in this investigation.

_____

_____

2. Use this table to record your data.

3. How do the results of this investigation compare to the results obtained when using foil? Explain any difference in the results.

_____

_____

_____

Time (min)	Black Paper Temp. (°C)
1	
2	
3	
4	
5	
6	
7	
8	
9	
10	

© Harcourt

Name _____

Date _____

Test the Shape and Size of Sails

1. Observe and Ask Questions

A boat race is called a regatta. When sailboats are being
raced, the speed of the boat has a lot to do with the size and
shape of the sails. What size and shape sail will catch the
most wind and help a sailboat win a regatta? Make a list of
questions you have about the sails on sailboats. Then circle
a question you want to investigate.

_____

2. Form a Hypothesis

Write a hypothesis. A hypothesis is a scientific explanation
that you can test.

_____

3. Plan an Experiment

Identify and Control Variables To plan your experiment,
you must first identify the important variables. Complete the
statements below.

The one variable I will change is

_____

The variable I will observe or measure is

_____

© Harcourt

The variables I will keep the same, or *control*, are

_____

Develop a Procedure and Gather Materials Write the steps you
will follow to set up an experiment and collect data.

_____

_____

_____

_____

_____

_____

_____

_____

_____

_____

_____

Use extra sheets of blank paper if you need to write down more steps.

Materials List Look carefully at all the steps of your procedure,
and list all the materials you will use. Be sure that your teacher
approves your plan and your materials list before you begin.

_____

Name ______________________

Inquiry Log

4. Conduct the Experiment

Gather and Record Data Follow your plan and collect data. Use the chart below or a chart you design to record your data. **Observe** carefully. **Record** your observations, and be sure to note anything unusual or unexpected.

Sail Number	Area of Sail (sq. cm)	Distance Ship Traveled (cm)
1		
2		
3		
4		
5		
6		
7		
8		
9		
10		
11		
12		

Other Observations

____

____

Interpret Data Make a graph of the data you have collected. Put the area of the sail on the *x*-axis and the distance the ship sailed on the *y*-axis. Plot the graph on a sheet of graph paper, or use a computer graphing program.

5. **Draw Conclusions and Communicate Results**

   Compare the hypothesis with the data and graph, and then answer these questions.

   1. Was your hypothesis supported by the results of the experiment? Explain.

   _____

   2. How would you revise the hypothesis? Explain.

   _____

   _____

   3. What else did you **observe** during the experiment?

   _____

   _____

Prepare a presentation for your classmates to **communicate** what you have learned. Display your data table and graph.

Investigate Further

Write another hypothesis that you might investigate.

_____

Name _____

Date _____

Light a Bulb

Materials

D-cell battery

30-cm length of insulated wire with stripped ends

miniature light bulb

masking tape

Procedure

CAUTION: Don't touch the sharp ends of the wire!

1. Using these materials, how can you make your bulb light? **Predict** what setup will make the bulb light.

2. Draw a picture of your **prediction**.

3. Test your **prediction**. Put the materials together the way your drawing shows.

4. **Record** your results. Beside your drawing, write *yes* if your bulb lit. Write *no* if it did not.

5. Make more **predictions** and drawings. Test them all. **Record** the results of each try. Draw the set up that works the best.

Name _____

Draw Conclusions

Inquiry Skill Tip

Clear records help you tell what you did during an activity. You can draw or write to **record data**. Keep your notes in order. Then someone can repeat what you did.

1. How must the materials be put together to make the bulb light?

_____

_____

2. **Inquiry Skill—*Record data*** Look at your drawings and notes. How did you **record data**? Scientists publish their results so that others can check them. Could someone use your records to double-check your tests? Explain.

_____

_____

_____

Investigate Self-Assessment	Agree	Not Sure	Disagree
I followed the directions for this Investigation.			
I was careful about touching the bare ends of each wire.			
I used my drawings and notes to **record data** about the correct way to light the bulb.			

© Harcourt

Name _____

Date _____

Independent Inquiry

Predict the kind of setup that will light two bulbs at the same time. Test your prediction.

Materials

- D-cell battery
- insulated electric wire
- 2 miniature light bulbs
- masking tape

1. Write your prediction.

   _____

   _____

2. Try it. Describe what happens.

   _____

   _____

3. How did the placement of wires differ when you were trying to light two bulbs instead of one?

   _____

   _____

4. Explain how the outcome of this investigation was the same or different from your prediction.

   _____

   _____

Can Electricity Make a Magnet?

Materials

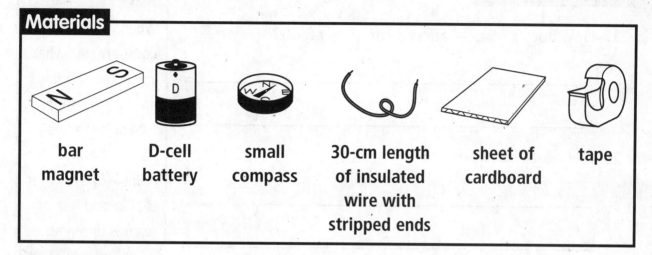

| bar magnet | D-cell battery | small compass | 30-cm length of insulated wire with stripped ends | sheet of cardboard | tape |

Procedure

1 Move a bar magnet around a compass. **Observe** what the compass needle does. Put the magnet away.

2 Tape the battery to the cardboard. Tape one end of the wire to the flat end of the battery. Leave the other end loose.

3 Tape the wire to the cardboard in a loop, as shown in the picture on page 619 of your textbook.

4 Place the compass on top of the loop. **Observe** the direction in which the compass needle points.

5 Touch the loose end of the wire to the pointed end of the battery. **Observe** what the compass needle does. **Record** your observations on the chart.

Bar Magnet	Battery

© Harcourt

Name _____

Draw Conclusions

1. How does a magnet affect a compass needle?

   _____

   _____

2. How does an electric current affect a compass needle?

   _____

   _____

Inquiry Skill Tip

When you **compare**, list what you know about each item. Use the lists to make two new lists about the items—one of what is alike and one of what is different.

3. **Inquiry Skill—*Compare* Compare** electricity and magnetism. How are they alike?

   _____

   _____

Investigate Self-Assessment	Agree	Not Sure	Disagree
I followed the directions for this Investigation.			
I noticed differences between magnetism and electricity by **comparing** their effects on the compass needle.			

© Harcourt

Name _____

Date _____

Independent Inquiry

Repeat Steps 4 and 5 with the compass *under* the wire.
Compare your observations.

Materials
■ **30-cm length of insulated wire with stripped ends**   ■ **bar magnet**
■ **a sheet of cardboard**   ■ **D-cell battery**
■ **small compass**   ■ **tape**

1. Write your prediction.

_____

_____

2. Try it. Describe what happens.

_____

_____

3. Explain how the outcome of this investigation was the same or different from your prediction.

_____

_____

4. Why do you suppose that the experiment had this outcome?

_____

_____

_____

© Harcourt

Name _____

Date _____

The Ups and Downs of Energy

Materials

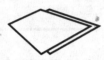

| strip of lightweight poster board, about 30 cm × 70 cm | books | ruler | marble | ball point pen | masking tape |

Procedure

1 Build a "roller coaster" track for your marble. Using your ruler, draw a line along both long edges of the poster board strip, about 1 cm in from each edge. Press hard with a ballpoint pen. Fold up along the marked lines to make walls. (This will keep the marble from rolling off the track.)

2 Place the strip between two stacks of books so that it forms a valley. Tape the ends of the strip to the books.

3 Hold a marble at the top on one side. Let go and observe. Does it go past the bottom and all the way up the other side? **Hypothesize** what affects the marble's path.

4 Based on your hypothesis, predict what will happen when you change your setup. Test your prediction. Draw your roller coaster below.

Our Roller Coaster

Draw Conclusions

1. What was the source of energy for the marble?

_____

_____

2. Inquiry Skill—*Change a Variable* To make a roller coaster that worked, you had to **change a variable**. What variable did you change? Explain.

_____

_____

Inquiry Skill Tip

When you experiment, you should change only one variable for each test.

To **identify and control variables**, list what you will change and what you will measure or observe. No other variables should change.

Investigate Self-Assessment	Agree	Not Sure	Disagree
I followed the directions for this Investigation.			
I **identified variables** that made a difference in my experiment.			

Name _____

Date _____

Independent Inquiry

Plan and conduct an investigation. Determine whether the weight of a marble affects the way the marble rolls.

Materials

- strip of lightweight poster board, about 30 cm × 70 cm
- ballpoint pen
- masking tape
- books
- ruler
- marbles

1. Write the plan for your investigation.

_____

_____

2. Try it. Describe what happens.

_____

_____

3. Explain how the outcome of this investigation was the same or different from your prediction.

_____

_____

4. Make a data table of your findings. In one column, write the size of the marble. In the other, record how the marble traveled. Make a graph of your findings.

© Harcourt

Name _____

Date _____

Energy Sources and Uses

Materials

colored construction paper
(5 yellow and 5 blue)

pens, pencils, drawing materials

Procedure

1 On each blue card, write *Uses of Energy*. On each yellow card, write *Sources of Energy*.

2 With your classmates, brainstorm a list of five ways you use energy every day. Draw a picture of each one on a separate blue card. Label each with a word or two.

3 Brainstorm places your energy comes from or ways you get energy. Make five yellow source cards.

4 Match a correct source card with a correct use card.

5 Mix your cards, exchange sets with a classmate, and work to match up the sources and the uses.

Uses of Energy	Sources of Energy

© Harcourt

Draw Conclusions

1. In what ways do you and your classmates use energy? What are the sources of the energy you use?

_____

_____

2. **Inquiry Skill—***Classify* Sort your cards to **classify** the uses and sources of energy you listed. Give reasons for how you sorted the cards.

_____

_____

_____

Investigate Self-Assessment	Agree	Not Sure	Disagree
I followed the directions for this Investigation.			
I came up with one or more ways to classify uses of energy, and sources of energy.			

© Harcourt

Name _____

Date _____

Independent Inquiry

**Make cards that show how energy moves and changes from
a source through one use of it. Communicate this sequence
to your classmates.**

Materials

- **10 large index cards, (5 yellow and 5 blue)**
- **pens, pencils, drawing materials**

1. What source and use did you choose to explain?

_____

_____

2. Write the steps between your source and your use.

_____

_____

3. Trade cards with a classmate, but don't show either your source or
your use cards. Based on the steps you were given, predict the source
and use.

_____

_____

4. Was your prediction correct or not? What clues did you use to make
your prediction?

_____

_____

Name _____

Date _____

Walk This Way

Materials

paper   pencil

Procedure

1 Choose a location in your school. The location could be an exit door or a bench, for example. A person going there from your classroom should have to make some turns.

2 Start walking to the selected location. As you walk, **record** the way you move. Include the distance you walk, where you turn, any landmarks you see, and how fast you move.

3 Return to the classroom. On a piece of paper, write directions to the location. Use your notes to add details. Don't name the location. Have a classmate follow the directions and try to match your speed.

4 Get feedback from your classmate about how well your directions worked. Use the feedback to improve your directions. Then follow the improved directions.

5 Switch roles with your partner, and repeat Steps 1–4. Record important information about good directions below.

Landmark	Direction of Movement

© Harcourt

Draw Conclusions

1. How did your partner know how far and how fast to walk and where to turn?

_____

_____

_____

2. **Inquiry Skill—*Communicate*** Tell why the revised directions and a good experiment plan are both examples of clear **communication**.

_____

_____

> ### Inquiry Skill Tip
>
> When you **communicate** you share ideas. Scientists communicate results of experiments. They also tell how they carried out the experiments so that others can check their results.

Investigate Self-Assessment	Agree	Not Sure	Disagree
I wrote directions to the location that I chose.			
I used the details to enhance directions along different parts of the path.			
I **compared** my directions to the procedure of an experiment.			

© Harcourt

Independent Inquiry

Draw a map to your location. Trade maps with a new partner. Compare using a map to using written directions.

Materials

Here are some materials that you might use.   ■ paper   ■ pencil
List additional materials that you need.

1. What will you investigate?

_____

_____

2. Use the table below to compare using a map and using written directions.

Ways They Are Alike	Ways They Are Different

3. What can you conclude from your investigation?

_____

_____

© Harcourt

Name _____

Date _____

Which Way the Ball Blows

Materials

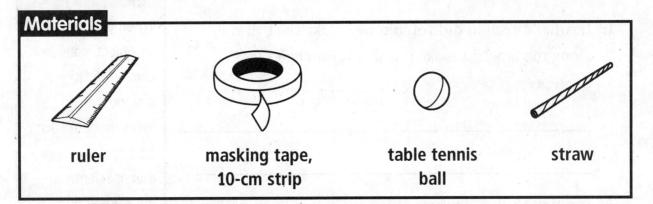

ruler   masking tape, 10-cm strip   table tennis ball   straw

Procedure

1 Put the strip of tape on a table or your desktop. Place the table tennis ball at one end of the tape.

2 Blow through the straw onto the ball. Blow gently and steadily. Make the ball roll along the tape. **Observe** whether the ball rolls in the direction in which you blow.

3 Place the table tennis ball back at the end of the tape. Blow on the ball at a right angle to the tape. **Observe** whether the ball rolls in the direction in which you blow.

4 Roll the ball gently. Blow on the ball in a direction different from the way the ball is rolling. **Observe** what happens.

	Observations
Step 2	
Step 3	
Step 4	

© Harcourt

Draw Conclusions

1. In what direction did you blow to make the ball roll along the tape? To make it roll at right angles to the tape?

   _____

   _____

2. **Inquiry Skill—*Measure*** What tools could you use to **measure** the motion of the ball?

   _____

   _____

Investigate Self-Assessment	Agree	Not Sure	Disagree
I followed the directions for this investigation.			
I **observed** the effects of blowing on the ball in various directions.			
Based on my observations, I **inferred** that the force of air directs the ball's movement.			

© Harcourt

Independent Inquiry

Blow on the ball in one direction. Have a partner blow at right angles to the direction in which you blow. Observe the path of the ball.

Materials

- ruler
- masking tape, 10-cm strip
- table tennis ball
- straws

1. Write your prediction for the path the ball will take when you blow the ball in one direction and your partner blows the ball at right angles to the direction you blow.

   _____

   _____

2. In the space below, draw a diagram of your investigation. After drawing the ball, draw and label arrows to represent the direction you blow, the direction your partner blows, and the direction the ball moves.

3. Did your results agree with your prediction? Explain.

   _____

   _____

   _____

Name _____

Date _____

Making Circular Motion

Materials

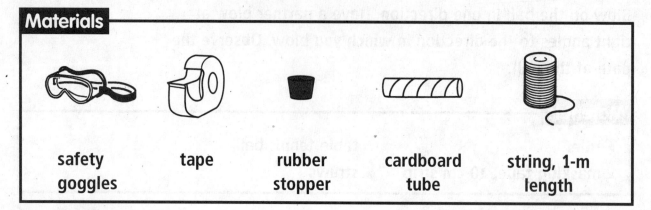

| safety goggles | tape | rubber stopper | cardboard tube | string, 1-m length |

Procedure

1 **CAUTION: Put on safety goggles.** Tie the stopper to the end of the string.

2 Tape over the string holding the stopper in place. Be sure the stopper is securely fastened. Thread the string through the cardboard tube.

3 Go outdoors. Stand so that you have a clear space 3 meters around you in any direction.

4 Holding the string, move the tube to whirl the stopper in a circular path over your head. Be careful to hold the string securely. You may want to wrap it once around your hand. Don't strike anything with the stopper! The circular path of the stopper should be level with the ground.

5 **Observe** the direction in which you are pulling the string as the stopper moves through the circle. Record your results.

Direction of Pull	Direction of Stopper

Draw Conclusions

1. In what direction did you pull the string to keep the stopper moving in a circle?

   _____

2. **Inquiry Skill—*Experiment*** What materials would you need for an **experiment** to test the hypothesis that spinning the stopper faster requires a stronger pull?

   _____

   _____

Inquiry Skill Tip

When you **experiment**, you carry out a fair test of a hypothesis. For a fair test, you need to be sure that you change only one variable. The other conditions should be controlled, or the same, for every test. Each time, you should test and gather data in the same way.

Investigate Self-Assessment	Agree	Not Sure	Disagree
I followed the directions for making circular motion with the string and stopper.			
I obeyed safety rules by wearing the safety goggles.			
I suggested materials for a possible **experiment**.			

© Harcourt

Independent Inquiry

Try whirling the stopper in an up-and-down circle in front of you. Observe the direction in which you pull. How is it different from the direction in Step 5?

Materials

- safety goggles
- tape
- rubber stopper
- string, 1-m length

1. In what direction do you predict you will have to pull the string?

_____

_____

2. Describe your observations in the investigation.

_____

_____

3. Do your observations agree with your prediction? Explain.

_____

_____

_____

Up and Down

Materials

safety goggles   2 rubber bands   tape   wooden ruler

Procedure

CAUTION: Put on safety goggles.

1 Work in groups of three. Slip a rubber band onto each end of the ruler. Tape the rubber bands in place on the bottom of the ruler, 2 cm from each end.

2 One person should hook a finger through each rubber band and lift the ruler. This person should keep the ruler level while a second person presses down on the 15-cm mark.

3 The third person should **measure** the length of each rubber band. **Record** your observations and measurements.

	Length of Rubber Band Closest to 0 cm (cm)	Length of Rubber Band Closest to 30 cm (cm)	Observations
15-cm mark			
17-cm mark			
19-cm mark			
21-cm mark			

© Harcourt

4 Repeat Steps 2 and 3 with the second person pressing on the 17-cm mark.

5 Repeat Steps 2 and 3 with the second person pressing on the 19-cm and 21-cm marks.

Draw Conclusions

1. What happened as the second person pressed farther from the ruler's center?

   _____

   _____

2. **Inquiry Skill—*Use Space Relationships*** Sometimes scientists can learn about what they can't see by watching how it affects other things. For example, the push on the ruler affected the rubber bands. How did you **use space relationships** to observe what was happening to the ruler and rubber bands?

   _____

   _____

> **Inquiry Skill Tip**
>
> When you **use space relationships**, you look at how objects move and take up space. Watch for how objects change size. Watch for how they change position compared to each other.

Investigate Self-Assessment	Agree	Not Sure	Disagree
I followed the directions for setting up the ruler and rubber bands.			
I obeyed safety rules by wearing goggles.			
I used **space relationships** to learn about forces.			

Investigate Log

Independent Inquiry

What do you predict would happen if you pressed down on the 10-cm mark? Test your prediction.

Materials

- safety goggles
- wooden ruler
- 2 rubber bands
- tape

1. Predict what will happen when you press down on the 10-cm mark. Use the results of the Investigate to justify your prediction.

   _____

   _____

   _____

2. Record your observations for the investigation.

   _____

   _____

3. Did the results support your prediction? Explain.

   _____

   _____

   _____

Name _____

Date _____

Hoist Away

Materials

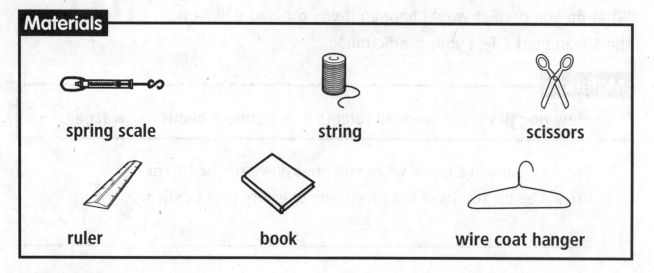

spring scale   string   scissors

ruler   book   wire coat hanger

Procedure

1 Work with a partner. Tie a loop of string tightly around the book.

2 Bend the hanger into a diamond shape. Hang it from a doorknob or coat hook.

3 Cut a 1-m length of string. Tie one end to the loop of string around the book. Pass the other end through the hanger, and attach it to the spring scale.

4 One partner should lift the book by pulling down on the spring scale. Note the reading on the spring scale. The other partner (continued)

	Step 4 (Pull Down)	Step 5 (Pull Up)
Spring scale reading		
Distance spring scale moved (cm)		
Distance book moved (cm)		
Observations		

© Harcourt

should measure the distance the spring scale moved
and the distance the book moved. **Record** your
observations and **measurements**.

5 Hook the hanger to the loop of string on the book.
Untie the long string. Pass it through the hanger, and
tie it to the doorknob or coat hook. Repeat Step 4,
but pull up on the spring scale.

Inquiry Skill Tip

For an experiment, be
sure that you **identify
the variables** that you
will control, or keep the
same. You also must
know which variable you
will change and which
one you will measure or
observe.

Draw Conclusions

1. How did the force needed to lift the book change?

_____

_____

2. Inquiry Skill—*Variable* What **variable** changed the second
time?

_____

_____

Investigate Self-Assessment	Agree	Not Sure	Disagree
I set up the string, book, hanger, and spring scale according to the instructions.			
I was careful when working with the wire coat hanger and the scissors.			
I **identified the variables** that were tested.			

© Harcourt

Name _____

Date _____

Independent Inquiry

How could you use two hangers with the spring scale and string to lift the book? Try it!

Materials
Here are some materials that you might use. List additional materials that you need.
■ spring scale ■ scissors ■ string
■ 2 wire coat hangers ■ book ■ ruler

1. Draw and label a picture to show how you will use two hangers with a spring scale and string to lift the book.

2. Make a data table, and record your measurements and observations for using this setup to lift the book.

3. Was your setup successful in supporting the book? Compare it to the setups used in the Investigate.

_____

_____

© Harcourt

Moving Up

Materials

tape measure

cardboard

scissors

spring scale

string

toy car

Procedure

1 Use some of the cardboard to make a ramp from the floor to a chair seat. Make a second ramp, twice as long as the first. Using the tape measure, find and **record** the distance from the floor to the seat, both straight up and along each ramp.

2 Tie a loop of string to the toy car. Attach the spring scale to the string.

3 Hold on to the spring scale, and lift the car from the floor directly to the chair seat. **Record** the force shown.

	Distance (cm)	Force Shown on Spring Scale (N)
Straight up		
Short ramp		
Long ramp		

© Harcourt

4 Hold on to the spring scale, and pull the car up the short ramp from the floor to the chair seat. **Record** the force shown. Do the same for the long ramp.

Draw Conclusions

1. How did using the ramps affect the amount of force needed to move the car to the chair seat?

_____

_____

2. **Inquiry Skill—***Interpret data* Scientists **interpret data** to draw conclusions. After examining your data, what conclusions can you draw?

_____

_____

_____

> **Inquiry Skill Tip**
>
> When you **interpret data**, you make a statement that agrees with all of the data and measurements in an investigation. Think about whether the measurements increase or decrease. Compare the measurements that you make to see if you notice any patterns or relationships among them.

Investigate Self-Assessment	Agree	Not Sure	Disagree
I built and set up the ramps according to the instructions.			
I used the tape measure to find the distance from the floor to the chair seat straight up and along both ramps.			
I **drew conclusions** about the investigation based on the data.			

© Harcourt

Name _____

Date _____

Independent Inquiry

Predict what variables affect the force needed to lift
the car. Plan and conduct a simple investigation to test
your ideas.

Materials

Here are some materials that you might use.
List additional materials that you need.

- tape measure
- toy car

- cardboard
- scissors

- spring scale
- string

1. List two other things that might affect the force needed to lift the
   car. Predict how each one will affect the force.

   _____

   _____

2. Choose one of the variables you listed above. In the space below,
   make a chart or diagram to show the steps you take to test its
   effect on the force needed to lift the car.

3. Describe the results of your investigation. Do they support
   your prediction?

   _____

   _____

Science Fair Project Ideas

Disinfectant Test

After you've learned about fungi, design and conduct an experiment to test how well various soaps and disinfectants prevent the growth of mold on fruit, such as apples or oranges. You will need to start a mold culture on some object and then use it to inoculate treated fruit samples. Remember to keep mold samples contained and dispose of them properly.

Camouflage Test

Conduct an experiment to test the effectiveness of color and pattern camouflage. Make various background "environments" and then give volunteers a limited amount of time to hunt for various colored or patterned objects on each background.

Day length and Plant growth

Design and conduct an experiment to find out how exposure to different periods of daylight affects the growth of one type of plant, such as corn, beans, or peas. Use inexpensive digital kitchen timers to remind yourself when to put your plants in a dark place.

Fitness and Heartbeat Rate

After you learn about the respiratory and circulatory systems, design and conduct an experiment to find out how regular exercise affects a resting heartbeat rate (pulse after sitting quietly for three minutes). Possible exercises include walking and stair-climbing. Talk to a parent or trusted adult about where and how to exercise safely.

Science Fair Project Ideas

Do Birds Have Color Preference?

Design an experiment to determine if birds have a color preference.
Milk cartons can be used to build three identical bird feeders. Cut
openings in each of the milk cartons. Use dowels to make perches in
each of the bird feeders. Paint one bird feeder white, one red, and
one blue. Place equal amounts and kinds of bird feed in each feeder.
Suspend them in a tree. You should record the number and variety of
birds that come to each feeder. Graph and display your results.

Protecting Ecosystems

Plan and conduct an experiment to test ways of reducing erosion. You
can explore how adding plants, ground coverings, or structures affects
erosion on a small area of sloping land. Make sure you obtain
permission from a responsible adult before you begin work.

Science Fair Project Ideas

Volcanoes

Plan and conduct an experiment to find out how thickness of lava affects the formation of a volcano. You can use food materials such as frosting, pudding, peanut butter, butter, and margarine. The materials can be diluted, warmed, or mixed with solids to model different thicknesses and "chunkinesses" of lava.

Fossils

Plan and conduct an experiment to find out which petroleum oil lubricates the best. You can use household oils and clean, unused motor oils. The force required to drag an object can be measured using a rubber band or a spring scale.

Science Fair Project Ideas

Weather or Not

It's time to calculate the percentage of accurate weather predictions! Collect the weather maps from your local newspaper. You can build a chart of actual weather for each day and compare the weather predictions with the actual weather. Calculate the percentage of accurate predictions and display your findings.

Reducing Drag in Oceans

Plan and conduct an experiment for the hypothesis that mammals living in the ocean have body adaptations to reduce drag as they swim. You should make shapes out of clay or wood. Then use a spring scale to measure the force needed to pull the shapes through water.

Sunset Times

Plan and conduct an experiment to find out how the length of the day changes. Carefully observe the time of sunset each day for a few weeks. Use your data to predict sunset times for the next week.

Science Fair Project Ideas

Thick Liquids

Design an experiment to find out how the thickness (viscosity) of a liquid affects how objects sink into it. You can drop fishing weights into equal amounts of liquids. Use a stopwatch to measure the time for the fishing weights to reach the bottom of each container.

Conducting Heat

Design an experiment to find out how long it takes heat to travel through a solid. You can use a metal skewer and attach small blobs of wax to it. With an adult's help and using a pot holder, you can hold the tip end near a heat source. Then, use a stopwatch to measure the time it takes for each piece of wax to melt and fall off.

Megaphone Ears?

After you have studied sound, measure how much further you are able to hear when your "ears" are enlarged. Using paper cup ears, you can measure the distance at which you can hear a ticking clock. Then design and test the best "ear" to capture sound.

Light

Plan and conduct an experiment to find out how well people can predict colors. You can use food coloring to tint water in clear bottles or glasses. Several objects can be selected, each a different color. Find out what color the object appears when seen through the tinted water, when lighted with a beam shone through the water, and when seen in white light. Ask others to look at the object through the tinted water and predict what color it will be in white light.

Science Fair Project Ideas

Cling-Free

Design an experiment to find the best treatment to reduce the cling caused by static electricity. You can begin by rubbing air-filled balloons with wool to generate static electricity. The amount of static electricity can be inferred by observing how much confetti a balloon will pick up. Rub or coat the balloon with various liquids to see how the liquids affect static cling.

Animal Speeds

Design an experiment to measure the speed of an animal such as an insect, a bird, or a family pet. For example, you might measure the time it takes for a beetle to crawl through a plastic tube. Be sure that you treat all animals humanely.

Pull a Panda

Read a book such as *How Do You Lift a Lion?* by Robert E. Wells. You can design an experiment to find the most efficient way to pull a heavy object, such as a panda, on a pallet. A spring scale can be used to measure force. Provide graphs that illustrate your findings. If you used a book, remember to credit that book.